Construction Management

Construction Management

Eugenio Pellicer
PhD, Associate Professor (Senior Lecturer),
Universitat Politècnica de València, Spain

Víctor Yepes
PhD, Associate Professor (Senior Lecturer),
Universitat Politècnica de València, Spain

José C. Teixeira
PhD, Associate Professor (Senior Lecturer),
Universidade do Minho, Portugal

Helder P. Moura
MSc, Regional Director, Estradas de Portugal, SA, Portugal

Joaquín Catalá
PhD, Professor, Universitat Politècnica de València, Spain

WILEY Blackwell

This edition first published 2014
© 2014 by John Wiley & Sons, Ltd

Registered Office
John Wiley & Sons, Ltd, The Atrium, Southern Gate, Chichester, West Sussex, PO19 8SQ,
United Kingdom.

Editorial Offices
9600 Garsington Road, Oxford, OX4 2DQ, United Kingdom.
The Atrium, Southern Gate, Chichester, West Sussex, PO19 8SQ, United Kingdom.

For details of our global editorial offices, for customer services and for information about how
to apply for permission to reuse the copyright material in this book please see our website at
www.wiley.com/wiley-blackwell.

The right of the author to be identified as the author of this work has been asserted in
accordance with the UK Copyright, Designs and Patents Act 1988.

Library of Congress Cataloging-in-Publication Data

Pellicer, Eugenio, 1965–
    Construction management / Eugenio Pellicer, Víctor Yepes, José C. Teixeira, Helder P. Moura,
Joaquín Catalá. – First edition.
        pages   cm
        Includes bibliographical references and index.
        ISBN 978-1-118-53957-6 (pbk.)
    1. Construction industry–Europe–Management–Textbooks.   I. Title.
        HD9715.E82P37 2013
        624.068–dc23
                                                                        2013019567
A catalogue record for this book is available from the British Library.

Wiley also publishes its books in a variety of electronic formats. Some content that appears in
print may not be available in electronic books.

Cover images: courtesy of Shutterstock
Cover design by Sophie Ford, His and Hers Design

Set in 10/12.5pt Minion by SPi Publisher Services, Pondicherry, India

1   2014

Contents

About the Authors

Eugenio Pellicer received his MSc degree from Stanford University, USA, and his PhD from the Universitat Politècnica de València, Spain, where he is currently working as an associate professor in construction project management, and he is also in charge of the MSc in Planning and Management in Civil Engineering. He has participated in quite a few international projects with other European and Latin-American universities.

Víctor Yepes is a civil engineer and associate professor at the Universitat Politècnica de València, where he received his PhD. He is currently involved in several research projects related to optimisation and life-cycle assessment of concrete structures. He is currently teaching courses in construction methods, innovation and quality management. He is also the academic director of the MSc in Concrete Materials and Structures.

José C. Teixeira graduated in civil engineering from Porto University, Portugal, and holds a PhD in construction management from Loughborough University of Technology (UK). He is currently an associate professor at the University of Minho, Portugal, lecturing on construction project management at the School of Engineering. He is currently involved in several international projects with other European and Latin-American universities.

Helder P. Moura holds a degree in Civil Engineering from Porto University and an MSc in Construction Management (Minho University, Portugal). He has had broad experience in the construction field for over 20 years, with a special interest in claims management. He has also skills as an expropriation expert and real estate evaluator. Currently he is a regional director in the management and operation of a Portuguese road concessionary.

Joaquín Catalá is the former director of the MSc in Occupational Risk Prevention and of the Department of Construction Engineering at the Universitat Politècnica de València, where he is currently lecturing as professor. He obtained his PhD at this same university. He has supervised many MSc and PhD theses related to construction management in general, and occupational health and safety management in particular.

Preface

Human beings build houses, roads, bridges, tunnels, ports, airports, factories, shopping centres, offices, warehouses and so on. Public and private organisations participate in the execution of these projects with the collaboration of architects, engineers, companies and financial entities, among others. The scope of the 'construction' term includes the primary sector (quarries), secondary sector (equipment and construction materials) and tertiary sector (engineering and architecture consulting companies), that is, industrial, commercial and service activities. Therefore, it includes private initiatives and the activities of public administrations.

Construction is set within a sensitive sociopolitical environment, affected by the need to protect people's fundamental rights, such as public health, homes, road safety and environmental integration. It is easy to understand the problems in establishing a set of economic activities in an ordered and coherent framework, taking place in an environment with many players from various sectors, with conflicting interests and branches towards many other economic sectors. In a broader sense, construction is an important motor for – and, at times, an obstacle to – economic growth.

In view of this scenario, this book is focused exclusively in the construction phase of that process. The contractor's point of view is chosen, even though the links with the owner are always taken into account. An envisioned outline of the management at the construction site is looked for, from the signing of the contract to the beginning of the operational phase.

Aiming to develop a useful and applied text for students in postgraduate construction programmes, five authors have worked together. They all have a wide experience in the construction industry. In fact, this book is based on their previous experience in several Leonardo da Vinci projects financed by the European Union, and a draft of this text was the product of one of these projects: 'Common learning outcomes for European managers in construction'. Four of the authors currently do their main work as professors in two different universities: Eugenio Pellicer, Víctor Yepes and Joaquín Catalá at the Universitat Politècnica de València, and José Teixeira at the Universidade do Minho; Helder Moura works for the Portuguese Highways Agency. Each author has led those chapters in which they can add most value to the work due to their professional and academic practice. Nevertheless, all of them have contributed to the whole.

Every chapter begins with educational objectives that enable the reader to know in advance its focus. Furthermore, it includes many references throughout the text in order to provide clues for postgraduate students on key issues that can allow them to deepen their knowledge. Other recommended books for further reading are also indicated at the end of each chapter. European examples, best practices and procedures are incorporated using 'boxes' in the text. Furthermore, figures and tables try to clarify the main concepts for the readers.

The book is organised into 15 chapters, with Chapter 1 introducing the construction industry in general. This chapter also gives data on construction in the European Union and presents the construction company, including some of its traditional organisational hierarchies, and the link between the firm and the construction site.

Chapters 2 and 3 examine the contractual documents and the different agents that are involved in the construction phase, including the ones associated with relational contracts. Administrative and technical documents of the design phase and tendering documentation are also analysed, as well as the main phases of the estimating process, during bid preparation. Chapter 4 introduces other interesting issues such as communications, negotiation, information flow, documentation and record keeping; it develops daily logs, reports, the diary and meetings, paying particular attention to building information modelling.

Chapters 5 to 8 explain basic issues related to the execution of works, such as: site setup and planning (5), machinery and equipment (6), productivity and performance (7), and quality management (8). Chapter 5 considers constraints of the site and the equipment, storage of materials, temporary facilities, auxiliary works, jobsite offices and jobsite security. Chapter 6 looks at the selection of machinery, the calculation of its cost and machinery maintenance. Chapter 7 presents a study of works, techniques of work measurement, equipment performance and productivity assessment; the concept of benchmarking is explained here too. Finally, Chapter 8 explores the quality management processes at the construction site, and also deals with innovation and knowledge management processes in the construction organisation, relating them to quality management.

Chapter 9 focuses on health and safety on the construction site. Taking into account the European Union directives, the general principles of prevention and the involved agents and their duties are explained. Site-specific safety plans and incidents during the execution of works are also considered from the point of view of the site manager as well as the owner. Chapter 10 then shows the environmental management of the construction site; the issue of sustainability is considered, and construction and demolition recycling is explained in detail. Chapter 11 analyses the supply chain management in construction, introducing the lean construction approach.

Chapter 12 is the longest in the book. It describes resources management, investigating the scope of activities, the assignment of resources to activities, their sequence, duration and monitoring. It develops the ideas of bar and network diagrams, cost of resources and cost control. Subjects such as the last planner system, the earned value method, value engineering and risk management are also explained.

Chapter 13 focuses on progress payment procedures, while Chapter 14 considers changes and claims during the construction phase. Finally, Chapter 15 describes the closeout process of the construction works and also the construction contract, analysing the commissioning procedures, handover and occupation; it also introduces the operation and maintenance manual and the as-built documents.

Our gratitude goes to our former, present and potential students who have helped us to develop, in different ways, the materials of this book. We want to acknowledge our families for their patience during the time we have spent preparing the book instead of enjoying their company. We would also appreciate any comments that might help us to improve the manuscript in future editions.

The authors
Valencia and Porto, February 2013

1 Organising Construction Processes in Construction Companies

1.1 Educational outcomes

This chapter briefly analyses the construction industry, highlighting construction companies as the key productive element of the whole process and describing the construction project as the final result of this process. By the end of this chapter readers will be able to:

- identify the life-cycle of the facility (building or civil engineering infrastructure)
- recognise the production by projects as the main driver in the facility life-cycle
- formulate the particularities of the construction industry, from a European perspective
- summarise the characteristic of the construction companies that work in this market.

1.2 The facility life-cycle

The construction industry is complex, based as it is within the three major sectors (raw materials, manufacturing and services). It is generally seen as being in an intermediate situation, between industrial activities and services. Also, construction activity is very complex, as a result of a series of characteristics that define its activity, procurement and organisation. Its ultimate purpose is to design and complete a series of products (or facilities) and their subsequent commissioning for use, either free or with the corresponding

Construction Management, First Edition. Eugenio Pellicer, Víctor Yepes, José C. Teixeira, Helder P. Moura and Joaquín Catalá.
© 2014 John Wiley & Sons, Ltd. Published 2014 by John Wiley & Sons, Ltd.

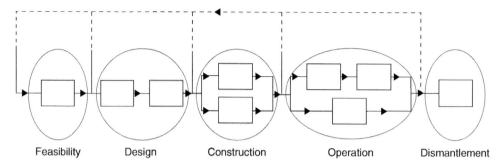

| Feasibility | Design | Construction | Operation | Dismantlement |

Figure 1.1   **Phases of the facility life-cycle and its different contracts.**

payment to a third party (Groàk, 1994). In the former case, the owner (client, promoter or developer) is public, and in the latter, the owner may be private (generally) or public (in some cases). In the context of this book, 'facility' is any building or civil engineering infrastructure (Wideman, 2003); the term 'infrastructure' will be also used synonymously through the text.

The construction industry can be viewed as a process that takes into consideration not only the execution (construction) of the facility, but also its feasibility analysis, design and operation (Groàk, 1994). This notion of process is equivalent to that of life-cycle (Levitt, 1965); it entails a set of phases starting with the initial concept and ending with the demolition of the facility, if necessary. Within the life-cycle approach, five main phases can be seen (Cleland, 1999): 1. feasibility, 2. design, 3. execution, 4. commissioning and operation – including maintenance – and, if required, 5. dismantlement. Different contracts can be present in each phase, either in series or in parallel (Figure 1.1). These contracts imply that the owners, as well as the companies involved, have to manage complex interfaces (Winch, 2010); an example of this would be the interpretation of the design by the contractor. The competitiveness of the companies working in this sector depends on the efficient operation of all the phases and interfaces. For example, an engineering consulting company may produce a good design but then it is not implemented for reasons beyond the control of the firm, or a construction site is permanently subject to stoppages because of poor management by the owner (licences, permits, etc.).

Feedback can happen over the whole life-cycle or in a part of it. To a large degree it will depend on the issues arising in the operation phase. At this phase maintenance is a key task, regardless of the operational approach and facility type. Anyway, it is important to realise that the quality and quantity of maintenance work will determine the service life of the facility. Once the regular life of the facility is over, three options are feasible (Figure 1.2): change of use, renovation or dismantlement.

The change of use of the facility is an option that focuses on a new approach to the operation phase. An example might be the owner of a building who decides to change its current rent option for selling the apartments to

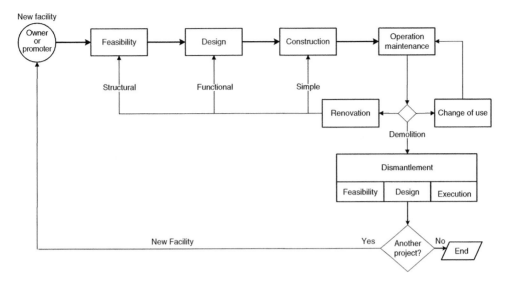

Figure 1.2 Detailed phases and feedback of the facility life-cycle (Pellicer et al., 2012) (Reproduced with permission from Colegio de Ingenieros de Caminos, Canales y Puertos).

customers (homebuyers). This involves a number of administrative and technical steps and actions in order to adapt the building to the new form of operation, but this change will not necessitate any construction works.

Renovation, with or without a change of use, can lead to many different situations. Three options are typical, depending on the nature of the renovation (Pellicer et al., 2012):

- **structural renovation:** when the structure (mechanical framework that holds the facility) is affected, a major intervention is required starting with feasibility analysis to ensure that the facility can evolve from its current state to the required one. This is the typical case of a bridge with a damaged foundation that needs to be repaired (or, alternatively, the bridge dismantled) or a building that needs the addition of an upper storey for a new use
- **functional renovation:** in this case a new design is needed. An example might be the addition of another carriageway on a highway without increasing the platform (it needs the adaptation of the lighting, signalling, marking, guardrails, etc) or the transformation of an old health centre into an office building
- **simple renovation:** refurbishments and repairs that do not modify the structure or design of the facility, such as changing guardrails and signalling in motorways, or changing small elements in buildings (doors, windows, bathroom remodelling, etc.).

Finally, dismantlement involves the demolition of the facility; this option can happen when the service life ends. The complexity of this phase requires a feasibility study in most of the cases, a demolition project and, finally, the implementation of the 'deconstruction'. Therefore, this alternative involves three sub-phases in the above scheme matching the three initial phases of the life-cycle of the facility (Pellicer et al., 2012).

Figure 1.2 summarises the whole process with their options. In this scheme, the whole sweep from the first idea to the dismantlement of the facility can be appreciated. After dismantlement, in the case of a building, there is then a vacant lot on which a new building can be built with the focus and goal that the owner deems appropriate.

1.3   Production by projects

Companies are organisations created by man and adapted to the environmental circumstances to attain specified objectives (economic profit, business survival and welfare of its employees) through the production of goods or services. Production can take place traditionally or by projects. Traditional production follows the common logical process: design, production, marketing and sale of the product. In production by projects, the traditional order is modified: first an idea is sold to the client, then the contract is signed and, finally, a unique product is developed, adapted to the client's changing requirements (Pellicer, 2007). A project can generally be seen as a temporary organisation established in order to create a unique product or service. Based on this idea, production by projects focuses on the business structure based on temporary teams, with members who collaborate to attain a common goal.

In traditional industry, projects are executed sporadically, modifying the common business organisational structures, which are not suitable for working with a project approach. Projects are composed of specific teams, independent of the normal production of the company. In this case, there is management *of* projects, but not management *by* projects. Companies that use a managerial approach by projects follow two types of processes: business and project-based (Gann and Salter, 2000). The first type of process is continuous and repetitive, developing routines as a result of the recurrence and frequency of business activities, increasing the formalisation, normalisation and economies of scale. Project-based processes entail operations that are not based on routine and are not very repetitive; this limits the improvement of processes, normalisation and economies of scale. However, this type of process facilitates the adaptation to the environment and innovation.

In the construction industry, projects are managed using business organisations that are prepared to work systematically with this approach (Gann and Salter, 2000). Projects involve the normal production of the company, matching orders or contracts executed for clients. Work teams are not stable. The management *of* projects (at the productive level) coexists with the management *by* projects (at the business level). Throughout the production process, there are different types of actors, depending on the phase, in addition to the companies that operate in diffuse coalitions with other organisations throughout the

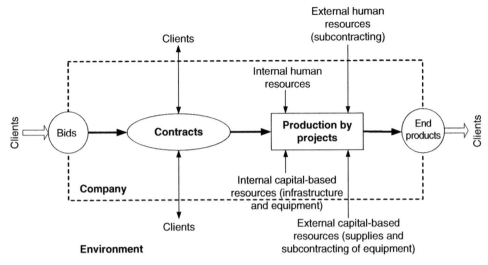

Figure 1.3   Business model (Pellicer et al., 2009) (Reproduced with permission from Pontificia Universidad Católica de Chile).

process (mainly in relations of owner–contractor–subcontractor). The feasibility and design phases are the basic field of action for engineering and architecture consultancies; in accordance with the degree of definition, the 'project' (in general) may consist of a feasibility study, a basic design project or a complete design project, among other elements. The construction phase is the main field of action of construction companies. In this case, the 'project' is represented by the actual construction works. Engineering and architecture consultancies can also participate in this phase through the inspection of the construction works acting as project managers on behalf of the owner. Other types of companies can appear throughout the construction process, with an important impact on other phases or on the whole process, such as, for example, developers (in private initiatives), service companies or concessionaires.

A typical business model is outlined in Figure 1.3 (Pellicer et al., 2009), and is based on an open systems approach in which the company interacts with the external environment. Each project originates in a bid prepared by the company to fulfil a request from a client. A signed contract between a company and a client follows, although this contract may be a verbal one. The aim of the contract can be any work that a company can perform during the construction process: feasibility studies, design documents, building infrastructures, maintenance works and so on. The production of the company is run by projects.

Resources are needed to deliver the end product. They may be internal (human resources, equipment and infrastructure) or external (supplies and subcontracting of human resources and equipment). It is likely that some of the company's resources will have to be transported from the central office to the work site in order to carry out a construction project. In addition, the firm is hierarchically organised, with well-defined personnel categories and work posts. This requires not only manpower, technicians and experts, but also administrative personnel. Moreover, every employee has a supervisor who oversees his/her work.

As stated previously, the activities that take place in a construction firm can be productive or administrative. The former are project-based, whereas the latter are business-focused (Gann and Salter, 2000). Administrative activities are non-productive activities, carried out mainly by administrative personnel who cannot be exclusively associated to a specific contract; consequently, they are calculated as overhead.

The construction companies follow a complex planning process, since production is discontinuous as a result of new contracts being constantly added. Strategic planning (long-term) depends on the political–economic cycle, with an irregular and hard-to-predict demand. Operational planning (short-term) is complex, since projects are subject to delays and changes caused by clients or third parties. These companies need to constantly adapt their organisation, in order to reallocate resources between different projects (Pellicer et al., 2009). The ideal objective is to achieve a continuous production, attaining the objectives, prioritising projects that provide an added value to the company and smoothing out the work peaks.

1.4   The construction industry

The purpose of this book is to describe the construction phase of the facility life-cycle, so the text will only focus on this phase from next section on. The productive activity of the construction phase is defined by the works execution, which presents specific peculiarities that condition the existence, structure and operation of companies working within this phase. The production starts after an order is placed. On the one hand, the product is unique: there are no two identical products, and the difference lies in the interaction with the terrain. In addition, production is temporary and intermittent: it has a start and an end. Finally, the construction procedures employed are not usually identical and, in many cases, can be subject to mechanisation. Other specific peculiarities of construction as a productive activity are (developed from Nam and Tatum, 1988):

- the final product involves a facility built on a specific site; the construction activity is carried out within the same place where the product will be set and not moved, which implies the spatial dispersion of the production process
- production is divided into many different parts. It can take place on any site of human activity, regardless of its importance, or even at any point of the world geography where a facility needs to be built. Construction sites constitute a relatively autonomous centre of work

- the finished product is extremely heterogeneous, given the large diversity of applications of construction products
- the size and complexity of the final facility is variable
- there are physical determining factors of the production process: topography, geology, use of natural resources, weather, urban planning, etc.
- the personality of the technicians taking part during first the design and then the construction has an impact on the final result.

The typical characteristics of the sector not only derive from the peculiarities of the final product and the production activity, but are also imposed through the market by demand factors (Pellicer, 2007). Demand from the private sector is materialised randomly in time and location. One of the immediate consequences is the low transparency of the market. In addition, as a consequence of its dispersion and division, there are large geographical fluctuations. The opacity of public demand is significantly lower, as a result of the legal requirements of publishing the tender prior to the awarding of the contract. In any case, the different bidders must compete in order to be awarded the contract. In most cases, the contract is awarded to the lowest bidder. As a consequence, the price of the product is formalised before the production takes place. This circumstance forces the entrepreneur to adjust the profit margins with a great precision. In some cases, particularly during slumps in the economy, bids can be so low that the entrepreneurial surplus can be zero or almost zero, where the only benefit would be ensuring their presence in the market for a period of time.

Therefore, it can be inferred that the sector is characterised by the production of heterogeneous and highly differentiated goods, which takes place in many different places and under different circumstances, with processes that are not usually subject to mechanisation, and working in many cases by order (Nam and Tatum, 1988). There is a strong correlation between economic cycles and production in the construction industry. During economic boom periods, the sector is one of the main driving agents of the economy, with indicators that are clearly above the mean, producing a pulling effect on the economy as a whole. But during depressions, the sector clearly drops below the average, especially regarding private investment. Therefore, public investments in infrastructure is a priority for public expenditure and a basic tool for the state's policy of boosting the economy during slumps; they also promote regional equilibrium, aiming at major social and economic objectives, including the stimulation of the creation of employment, favouring economic and social development. This is usually a key factor in the cohesion funds provided by the European Union. For example, Spain has directed the greatest part of the European Union's cohesion funds during the past few years to public infrastructure, mainly roads, railways and wastewater treatment plants. When the funds decreased (along with the increasing financial and property crisis), the Spanish construction industry lost energy; a similar situation happened to the UK, even though in this country the bubble was not so large. Nonetheless,

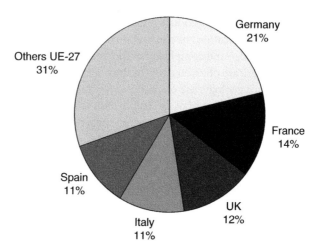

Figure 1.4   Production in construction of the EU-27, year 2011, by country (based on primary data from FIEC, 2012b).

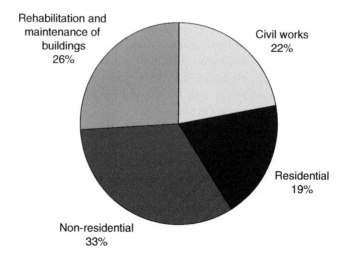

Figure 1.5   Production in construction in the EU-27, year 2011, by subsector (based on primary data from FIEC, 2012b).

as can be seen in Figure 1.4, there are still five countries (Germany, France, Italy, the UK and Spain) that hold approximately 70% of the production in construction in 2011 for the EU-27 (data obtained from FIEC, 2012b).

In accordance with data provided by the European Construction Industry Federation (FIEC, 2012a, b) regarding data from 2011 for the European construction industry (EU-27), the investment was €1200 billion, the average weight compared to the GDP was 9.6%, and employment comprised 7.0% of the total. The average investment in civil works was 22%, while the rest (78%) was concentrated on buildings, being 33% of the total for maintenance and refurbishment of these, as shown in Figure 1.5 (data obtained from FIEC, 2012b).

The current state of the European construction industry is generally stagnant, because of the economic crisis that is mainly affecting the southern European countries (FIEC, 2012a). In addition to the general characteristics of the construction industry, residential building (also referred to as the housing market) has some specific properties:

- the market includes not only the acquisition of homes, but also rental
- prices cannot be easily compared
- demand and supply operate with small delays in time
- the acquisition of a home generally depends on a mortgage loan to finance it.

Production in the construction industry does not have a high industrialisation capacity (and it cannot be easily replicated), despite the massive use of prefabricated materials. Final assembly is carried out within a specific location, with different topographical, weather, geological, hydrological, social and cultural characteristics. As already mentioned, demand determines prices, in accordance with local circumstances and the degree of entrepreneurial concurrence, which causes the variability in prices in the different local markets (Winch, 2010). Nevertheless, there is a correlation between the construction prices and income per capita, since costs – and especially manpower costs – are higher in developed countries. The construction sector in each country is also influenced by issues that offer a different analysis per nation: cultural habits, unreliable data, the predominance of small and medium companies, the informal economy, the legal and illegal migration and the high degree of subcontracting.

1.5 Construction companies

Construction companies transform the design into a facility ready for users. The production of these companies is the construction of a facility using a series of specific processes and procedures. The construction project starts with an order or contract with the client; it can be public (common in civil engineering) or private (more frequent in building). The execution of the construction works is based on a design project. The setup of the construction project is in the same site where the facility will be located. Therefore, a construction company is disseminated throughout the different construction sites where the company has construction projects under way. Construction companies coordinate the set of specialised suppliers and subcontractors with the purpose of executing the works. The supplies or inputs required are: labour force (own and subcontracted), equipment and machinery (own and subcontracted), materials, vehicles, offices and so on.

Currently, there is a strong trend towards diversification. Construction companies are frequently involved in the maintenance and operation of existing infrastructures; they also act as concessionaires and public-private

---

> ### Box 1.1 What are the main reasons for a contractor to internationalise its business?
>
> 1. Related to need:
>   a. The internal market is stagnant.
>   b. The competition among companies is very high.
>   c. There are unused resources in the company.
> 2. Related to opportunity:
>   a. International relationships offer timely actions to the company, either alone or in joint ventures with other partners.
>   b. Key clients go abroad and require support.
>   c. Foreign markets need specialised companies.
> 3. Related to strategic vision:
>   a. Entrepreneurial spirit and business vision.
>   b. Acquiring prestige and improving the image of the company.
>   c. Growing in size (manpower, contracts and turnover).
>   d. Reducing the risk by diversifying markets.
>   e. Increasing the profitability.

consortiums that have made a big impact on the market, through PFI/PPP contracts (Winch, 2010). Apart from the concession business, during recent decades, European construction companies have expanded their traditional business in order to include the management of utility services in general and infrastructures, in particular. These include waste management, urban furniture, sewage maintenance, public gardening, parking facilities and technical inspection of vehicles. This way, these companies have diversified their risks (out of the construction field) and invested their profits into businesses that allow increased synergies.

The next step (or a parallel step depending on the company and country) was the implementation of the main business (construction) in foreign countries – both developing and developed – with a minimum judicial security (Ofori, 2003). Consequently, foreign income represents an important percentage of their final turnover. Two factors have driven these companies to begin a process of international expansion. In the first place, construction has unfavourable characteristics from the point of view of risk: it is a very cyclical business with projects that take a long time to mature and the risk of cost increases, placing their financial viability in danger. The second factor is the strength that companies in the sector have acquired through decades of fierce competition, which shows a great capacity for managing their businesses. This international expansion is possible because of their in-depth knowledge of managing very complex projects (Bon and Crosthwaite, 2000). It meets several targets (Box 1.1), the main ones being: to reduce the risk of a cyclical business such as construction; to obtain additional contracts and turnover

> **Box 1.2   What are the strategies for a contractor to start working in a foreign country?**
>
> 1. From the point of view of partnering, the company could:
>   a. perform as a subcontractor hired by a main contractor from the same country of origin
>   b. act alone as a main contractor
>   c. form joint ventures with other companies.
> 2. From the point of view of the in-depth integration in the country, the company could:
>   a. bid for construction projects from its country of origin
>   b. create a branch or subsidiary firm in the foreign country
>   c. buy a local company in the foreign country.

(specially needed for listed companies) and to increase the profitability of their businesses.

Furthermore, each contractor can adopt several strategies for starting to work in a foreign country (Box 1.2), depending on the reasons for its internationalisation (internal analysis), the opportunities detected in the foreign country (external analysis) and the resources to hand. The case study presented in Box 1.3 can be also analysed as a real example of implementation.

The typology of these companies is very varied. The most common classification takes into account the size, in accordance with their production, but classification according to the number of employees can be misleading, due to the high level of subcontracting existing in the sector. The area of operation of the company can be decisive; there are companies that operate in local, regional, national or international markets. They can also be classified according to their specialisation. A common major division is between civil engineering and building; another is the type of client (public or private). Usually, large companies operate in national and international markets and offer wide-ranging services. Table 1.1 shows the top 15 contractors in the world (data from 2011). The first company (Hochtief AG) has belonged, since June 2011, to Grupo ACS (ranked second); thus this group has more than three times the revenues of the third company (Vinci).

Regarding the structure of business costs, 30% of the cost relates to human resources (regardless of the percentage of subcontracting), although technicians represent less than 5% of the company's costs. Of the production, more than 80% comes from the supply of materials and equipment, and the subcontracting of machinery and manpower. Billing must be done every month (generally through unit prices or lump sum contracts). The rotation of capital could even finance a company (Pellicer, 2007): 1. work is carried out; 2. the company's human resources are paid; 3. invoices are issued; 4. payment is collected; and 5. subcontractors are remunerated. The relative profit varies between 1 and 5%; even though the relative value seems low, the absolute value is high.

Box 1.3   Skanska, a case study of continuous adaptation to the market

Skanska is currently the ninth contractor in the world according to 2011 revenues (ENR, 2012). The company's mission is to develop, build and maintain the physical environment for living, traveling and working, whereas its vision is to be a leader in its home markets – the customer's first choice – in construction and project development (Skanska, 2012). The history of the company's earlier years is linked to Rudolf Berg, a chemical engineer who was appointed president of the cement producer *Skånska Cement*, founded in 1873 in Limhamn (Sweden). At that time, cement was mostly used for bricklaying. However, Berg perceived the huge potential for this material and developed a new casting technique. In 1887, he founded *Aktiebolaget Skånska Cementgjuteriet* (Skanska Precast Cement) in Malmö. The original purpose of the firm was the manufacture of cement-based decorative building elements, used for decorating Sweden's public buildings. Under the engineering vision of Berg, the company moved into the production of construction materials using concrete, mainly concrete pipes and floor plates. During that time, the French engineer François Hennebique (1892) patented the modern reinforced-concrete method of construction, and Berg took advantage of the system specialising the company in concrete structures. In 1897, Skanska had its first international contract: the supply of 100 km of hollow concrete blocks used for supporting telephone cables; the client was the UK's National Telephone Company. Yet in the 19th century, Berg's vision was to conduct a globalised business...

The change of century brought Skanska a huge contract, replacing the sewer system of St Petersburg, using concrete pipes. Throughout the following decades the firm contributed greatly to the development of infrastructure in Norway, Denmark, Finland and Sweden: roads, power plants, commercial and residential buildings. The company finished the Sandö Bridge in 1943: a 264 m span that was the world's longest concrete-based arch-bridge until the 1960s. The company also developed the Allbetong method, prefabricating elements for large-scale building projects; these elements were manufactured in Skanska's factories and put into place using construction cranes.

During the 1950s Skanska expanded internationally, moving to South America, Africa and Asia. At the beginning of the 1970s the company started to work in Poland, the Soviet Union and the USA. During the 1990s, Skanska had become one of the top US contractors, being involved in a large share of the Boston's Central Tunnel. Skanska was also a major shareholder in the Sundlink Consortium which won the contract to construct the Öresund Bridge and tunnel, linking Sweden with Denmark.

(Continued)

After 125 years of existence, the company is targeting the US and UK public–private partnerships (PPP) segment. It is well-established in the Scandinavian countries, the Czech Republic, Slovakia, the UK and Poland, covering the construction and investment businesses. In Latin America, Skanska is active in the oil, gas and energy sector and in PPPs. Currently, Skanska's operations are divided into four business units (Skanska, 2012): 1. construction, including residential building, non-residential building and infrastructures; 2. residential development of housing for sale directly to consumers; 3. commercial property development for leasing and divestment and 4. infrastructure development and investment in public-private partnerships (PPP).

Table 1.1   Top 15 international contractors considering only projects outside their home countries (based on primary data from ENR, 2012).

Rank	Firm	Revenue $ mil.
1	Hochtief AG, Essen, Germany	31 871
2	Grupo ACS, Madrid, Spain	31 148
3	Vinci, Rueil-Malmaison, France	18 674
4	Strabag SE, Vienna, Austria	17 289
5	Bechtel, San Francisco, Calif., USA	16 700
6	Saipem, San Donato Milanese (Milan), Italy	14 110
7	Fluor Corp., Irving, Texas, USA	13 527
8	Bouygues, Paris, France	12 608
9	Skanska AB, Solna, Sweden	12 339
10	China Communications Construction Group Ltd., Beijing, China	9 547
11	Technip, Paris, France	9 313
12	FCC Fomento de Construcciones y Contratas SA, Madrid, Spain	8 570
13	Construtora Norberto Odebrecht, Sao Paulo, SP, Brazil	7 351
14	Bilfinger Berger SE, Mannheim, Germany	7 146
15	Samsung Engineering Co. Ltd., Seoul, S. Korea	5 907

1.6   Organisational structure of a construction company

A company has to be managed effectively and efficiently in order to reach objectives planned by the upper management. This process is developed according to the procedures and policies previously set by the company. However, the main function of the company is production; and for a contractor, production means the construction of facilities or infrastructure (building or civil engineering). This process takes place within an organisational structure (Griffith and Watson, 2004). The organisation is the framework within which the company's tasks are divided, grouped and coordinated. The design of this structure for a construction company involves a decision-making process focused on the specialisation of tasks, chain of command, amplitude of control, decentralisation, formalisation and departmentalisation (Robbins

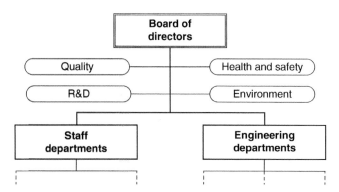

Figure 1.6 General organisation chart of a construction company.

and Coulter, 2010). The chain of command is the uninterrupted line of authority flowing from the highest levels to the lowest levels of the organisation. The organisation's hierarchy is specific, involving three fundamental concepts: authority, responsibility and unitary command. The hierarchical definition of each position is fundamental for the company's staff to inform and issue orders, and for employees to receive information and orders. The amplitude, control and consequently the number of levels of authority, must be sufficient to ensure a straightforward organisational structure, which can be flexible enough to adjust to the changing conditions of the market. The growth of the company involves a higher delegation of authority and responsibility.

Departmentalisation is the base on which tasks are grouped, in order to attain the organisation's objectives. Each organisation has its own way of classifying and grouping activities. Departmentalisation can be achieved by functions, products, geographical location or clients (Robbins and Coulter, 2010). Construction companies with a national or international scope usually have a geographical departmentalisation originating from business strategy and proximity to production issues. Within each geographic department (or branch) there can be a division by functions or products. Often, these branches are almost fully autonomous, depending on the central office for making critical decisions or for the generation of corporate policies. Smaller companies, with a local or regional character, have structures that are almost flat, whereby all employees report to the management. The degree of decentralisation, delegation and formalisation increases when companies grow in size.

The departmental structure is shown in the organisational chart of a construction company, in Figure 1.6. The organisational chart must not be simply seen as a graphical representation, but must also define the positions with their hierarchical relationships, establishing the objectives, functions, responsibilities and tasks. Usually, an organisational chart simply defines the management areas, line of support or 'staff' services and operational or 'engineering' line. The staff services have assistance functions: administration, ICT, clerical and so on. Line services have command functions covering personnel and technical functions over the company's productive activities.

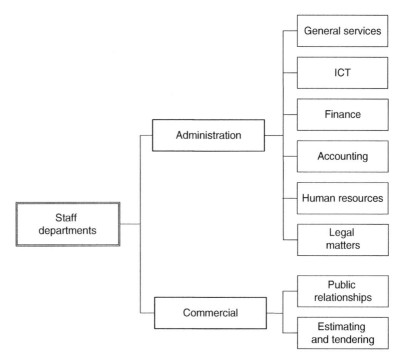

Figure 1.7   Organisation chart of the staff departments of a construction company.

Safety and health, environmental management, quality assurance and, recently, innovation (sometimes also including research and development) usually have their own areas, separated from the remaining departments of the company but directly linked to the board of directors (Harris et al., 2006).

In a construction company, the staff line is often divided in two independent departments: administration and commercial (Figure 1.7). The commercial department is very important in a construction company. They are in charge of tendering, bidding and public relations. The tendering process is basic in order to obtain new contracts to guarantee the permanence of the company in business. Usually, there is a period of one to two months to prepare a bid; the final economic proposal is established by the management area, depending on the business strategy (Harris et al., 2006). In general, although this might depend in the region, type of works and circumstances of the market, it could be considered 10% as the optimum awarding percentage.

The administration department includes: general services (reception, courier, photocopying …), ICT (information and communications technologies), finance, accounting, human resources and legal matters, among other areas. Currently, as a result of the operational decentralisation of medium and large companies, the adequate use of ICT is critical (smart-phones, laptops, tablets, iPads, email, internet, intranet, extranet, etc.); they can facilitate the transmission of information between the headquarters, branches and each of the

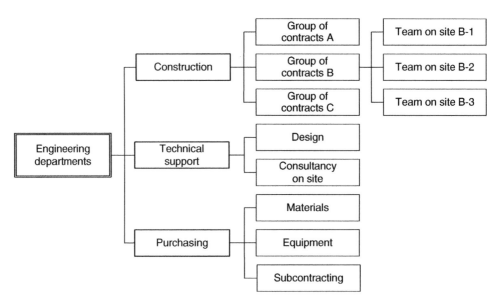

Figure 1.8 Organisational chart of the engineering departments of a construction company.

current construction sites. Therefore, a specific service is required within the company to manage the purchase and use of these basic tools; the ICT department is usually responsible for such tasks (Harris et al., 2006).

The engineering or operational macro-department is usually divided in three departments: construction, technical support and purchasing (including subcontracting). The construction department is fundamental in these companies, since it groups the works teams in accordance with their geographical area, type of works, client and so on. (Figure 1.8). After the awarding of the contract, the team responsible for its execution is emancipated from the company, establishing its own independent organisation which is used to execute the construction project. The real value of the company is mainly generated by the personnel working at the construction site.

The technical support services are usually located at the headquarters. They are used to offer their help in areas related to the design and detailed calculation of some elements, and to draft (totally or partially) 'as built' projects.

The purchasing department has an importance that depends on the degree of subcontracting of the company, as well as the extent and management of the company's own equipment. In some companies, it works as an independent business unit, renting the equipment internally and to third parties. In others, it operates as an internal service department, supplying the equipment required for each contract. In other cases, the company subcontracts the machinery supply to third parties for each project, thus avoiding maintenance and the internal management of the plant (Harris et al., 2006).

In large construction corporations, the general organisational chart presented in Figures 1.6, 1.7 and 1.8 can be replicated for each regional branch.

Another frequent option is the creation of intermediate organisational structures that maintain the centralisation of some staff departments (finance, personnel, legal, etc.) or the areas assisting the board of directors (quality control, health and safety, innovation, etc.).

1.7   The construction site within the construction company

As explained, during the construction phase, the design project has to be executed with the purpose of transforming it into an infrastructure that can be finally used. Therefore, the owner (client, promoter or developer) issues its orders to the construction company (or contractor) during this phase, generally through contracts. There is generally a representative of the owner for each contract (or construction project) that has administrative functions as well as inspection functions, depending a great deal on the type of contract and context (Fisk and Reynolds, 2006); this agent is called the resident project representative (USA), project manager (UK), facultative director (Spain), supervisor (Chile) or inspector (Colombia), depending on the country. The construction company manages different construction projects, on different sites and with a temporary limitation, all within the same business structure. The business organisation must take into account this plurality of production centres.

Each construction site manages its own human resources and materials. This organisation is simpler and more flexible than that of the company. Again, geographical dispersion and the immobility of the infrastructure is essential, and leads to different logistical problems at the construction site, regardless of the type of infrastructure or contractual conditions. The construction project requires a maximum degree of centralisation, as opposed to the decentralisation that is so common within companies. To sum up, the organisational structure at the construction site must be simple, with clear lines of dependence, while being flexible in order to adapt properly to the environmental circumstances (Pellicer et al., 2012).

A key element of the construction project is the construction site manager who performs as representative of the contractor (Dulaimi et al., 2005). This position is filled by a qualified technician in charge of the administration of the construction project, assuming full responsibility for planning, organisation, leading and control. This work post is generally filled by a university graduate. In addition, the construction site manager acts as the link between the company's macro-organisation and the works micro-organisation. The construction site manager is the last management step of the corporate ladder, but they are the top managing authority at the construction site (Figure 1.9). Under this position, there is a reduced team of technicians whose responsibilities include the execution, quality, risks prevention and surveying, among others.

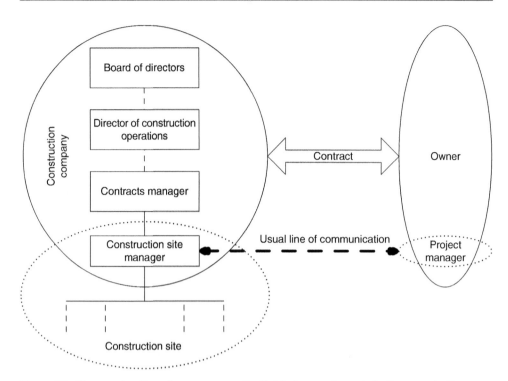

Figure 1.9 The construction site manager as the link between elements in the organisation.

In most cases, clients, collaborators, subcontractors, final users and even competitors look at the company in terms of the individuals who are managing the construction project (Winch, 2010). Thus, it is important to know how to transmit a good image of the company from the basic production centre represented by the construction site. The construction site manager is the driving agent of corporate culture, since they operate as the essential contact with the client, subcontractors and the social environment in general. If they are successful in their mission, they can be promoted up the corporate ladder. In most cases, the executives of construction companies are technicians who were promoted after spending much time and gaining experience through many construction projects.

References

Bon, R. and Crosthwaite, D. (2000). *The Future of International Construction.* Thomas Telford, London.

Cleland, I. (1999). *Project Management. Strategic Design and Implementation.* McGraw-Hill, New York.

ENR (2012). *The Top 225 International Contractors.* Engineering News Records, available 21/11/2012 at http://enr.construction.com/toplists/Top-International-Contractors/001-100.asp.

Dulaimi, M.F., Nepal, M.P., and Park, M.S. (2005). A hierarchical structural model of assessing innovation and project performance. *Construction Management and Economics*, 23(6), 565–577.

FIEC (2012a). *Annual Report 2011*. European Construction Industry Federation, Paris.

FIEC (2012b). *Key Figures – Activity 2011 – Construction in Europe*. European Construction Industry Federation, Paris.

Fisk, E.R. and Reynolds, W.D. (2006). *Construction Project Administration*. Pearson, New York.

Gann, D.M. and Salter, A.J. (2000). Innovation in project-based, service-enhanced firms: The construction of complex products and systems. *Research Policy*, 29(7), 955–972.

Griffith, A. and Watson, P. (2004). *Construction Management. Principles and Practices*. Palgrave Macmillan, New York.

Groàk, S. (1994). Is construction an industry? *Construction Management and Economics*, 12, 287–293.

Harris, F., McCaffer, R. and Edum-Fotwe, F. (2006). *Modern Construction Management* (6th Edition). Blackwell, Oxford.

Levitt, T. (1965). Exploit the product life-cycle. *Harvard Business Review*, 43(6), 81–94.

Nam, C.H. and Tatum, C.B. (1988). Major characteristics of constructed products and resulting limitations of construction technology. *Construction Management and Economics*, 6, 133–148.

Ofori, G. (2003). Frameworks for analysing international construction. *Construction Management and Economics*, 21, 379–391.

Pellicer, E. (2007). Consulting engineering companies versus building contractors: two different means of adapting to the market. *Revista de Obras Públicas*, 3483, 7–18.

Pellicer, E., Al-Shubbak, A. and Catalá, J. (2012). Towards a systematic vision of the infrastructure life-cycle. *Revista de Obras Públicas*, 3532, 41–48.

Pellicer, E., Pellicer, T.M. and Catalá, J. (2009). An integrated control system for SMEs in the construction industry. *Revista de la Construcción*, 8(2), 4–17.

Robbins, S.P. and Coulter, M. (2010). *Management* (11th edition). Pearson, New York.

Skanska (2012). *Annual Report 2011*. Skanska AB, Solna, Sweden.

Wideman, M. (2003). *The Role of the Project Life-cycle (Life Span) in Project Management*. AEW Services, Vancouver.

Winch, G.M. (2010). *Managing Construction Projects* (2nd Edition). Wiley, Oxford.

Further reading

Bon, R., and Crosthwaite, D. (2000). *The Future of International Construction*. Thomas Telford, London.

Winch, G.M. (2010). *Managing Construction Projects*. (2nd Edition). Wiley, Oxford.

2 Contract Documents

2.1 Educational outcomes

The aim of this chapter is to present a practical guide to the types of contract documents normally used in the construction industry. By the end of this chapter readers will be able to:

- comprehend and highlight the importance of the contract documents in the implementation of a construction project
- understand the meaning of the bidding documents
- examine the relevance of the contract agreement in order to choose the best applicable conditions and standard forms of the contract
- analyse the contract documents as well as the relevant project documents (bill of quantities, specifications and drawings).

2.2 Contract documents

To do business in construction, as well as in other commercial activities, it is necessary to establish a contract (Ashworth, 2012). In fact, any entity, public or private, that does not possess in-house capability or the methods and the means to construct, demolish, repair or maintain a built facility, should contract the execution of those works, or both the design and the execution. Thus, a construction contract can be defined as a legally binding agreement between two parties by which some rights and obligations are acquired by one or more acts, which are able to be enforceable in the courts, if it becomes necessary.

Construction Management, First Edition. Eugenio Pellicer, Víctor Yepes, José C. Teixeira, Helder P. Moura and Joaquín Catalá.
© 2014 John Wiley & Sons, Ltd. Published 2014 by John Wiley & Sons, Ltd.

Traditionally, the parties in a construction contract are: 1. the client, who wants to have some facility designed, constructed, maintained, repaired or demolished, and who has the necessary funds to pay for it; and 2. the contractor, who possesses the means and the methods to build that facility, for the agreed payment, according to the rules, specifications and schedules established in the contract documents. However, as well as the traditional type of the contract, there are several other forms of contractual arrangements between the client and the main contractor, owing to the many different procurement methods that can be used in construction, as explained in Chapter 3. Depending on the type of the project coalitions, different types of contracts are needed to determine rights and obligations, where the main parties (client and contractor) can play different roles. This is the case of specialised trader contractors, design-built contracting, different forms of public–private partnerships (long-term concessions, design-build-operate, finance and maintain), construction management, turnkey contractor, partnering contracts and so on. And for each one of these contracts, different documents apply (Ashworth, 2012).

The construction business is complex and many other parties are as important as the main contractor and the client, during one or more phases of the cycle (Chinyio and Olomolaiye, 2010): the design team (architects, engineers, quantity surveyors, etc.), the project management team (in the different forms of managing), specialised consultants of the client or of the contractor, different subcontractors, vendors, material and equipment suppliers, financing banks, funding entities, inspection and supervision on site (quality assurance, health and safety, environmental compliance and schedule accomplishment), equity detainers in the case of a concession, other contractors, future operators and maintainers of the constructed facility, and so on. In order to establish the commercial relationships between these internal stakeholders of the construction project (to understand the distinction between internal and external stakeholders, see Moura and Teixeira, 2010) different contracts are signed, each one with specific clauses and supporting documents.

The type of contracts vary from the simplest sell and buy (material, equipment), lending, lease or renting contracts (temporary site offices, for instance) to different types of managing contracts or services provision (specialised consulting, supervision, medical and nursing), working contracts (labour force and staff personnel), financing or factoring contracts, or back-to-back subcontractor contracts, to the more complex contracts such as the institutionalised public-private partnerships (PPP) contracts (including all forms of joint ventures between public and private stakeholders) or the most common contractual PPP. The last type of contracts includes: 1. the classic concession model, where the private partner has the obligation to design, build, finance and operate a public service/infrastructure (or just some of these functions) for a certain period of time and it is directly paid by the user of that facility; and 2. the private finance initiative (PFI), mostly used in social infrastructures where, for practical reasons, the private partner is paid by a public authority through fixed or variable rents and not by private users (examples are public lighting, hospitals, prisons, schools, roads with shadow tolls, etc.). Note that concession or PFI contracts, can take several forms, depending on the aim of the concession, namely concession of public works, concession of public services, continuous supply and provision of services, management of public facilities or just a collaboration contracts between partners (Li and Akintoye, 2003). And each one has different objectives, risks allocation, general and particular clauses and specific rights and obligations of the contracting parties.

Since the private construction contract is a particular form of a commercial contract, it does not require a special formality (or procedure) (that assumption depends on each country's legal systems), meaning that it can be substituted by a simple exchange of letters between the parties containing the conditions, or even by an oral agreement. A general rule for private contracts is freedom in its form and substance (Lowe and Leiringer, 2006), since first comes the practice and later the codification for the most common ones. In this case, every clause is permitted, given that it does not oppose general laws. However, in the situations where public entities are parties, or public funds are involved, a public contract applies, which is based on standard forms, applicable rules and clauses that are normally mandatory.

However, it is convenient for all the construction contracts, private or public, even for the simplest and smallest projects, to assume a written format, and use construction industry standards for general and particular forms of contracts, among other things to help minimise the occurrence of later disputes (Ashworth, 2012). In codified system countries, these requirements are generally included in civil law, when construction contracts are established between private actors, or in administrative law if a public entity is part of the contract.

But even with all applicable standard forms of contract (that will be discussed later in this chapter), for large complex projects such as nuclear power plants, main airports, oil refineries and similar facilities where the consequences of some risks are extreme, and which are only constructed once in a lifetime, the lack of a stable and durable relationship between client and contractor frequently led to major disputes. In these situations other forms of contracts can be used, namely relational contracting such as partnering, alliancing or the integrated project delivery (Chapter 3).

Whatever the type of contract, whenever a client needs to procure construction services, there are typically four different approaches:

- construction using in-house capability
- direct appointment of a desired contractor/supplier/entity
- issuing an invitation to a limited number of reliable contractors/suppliers
- opening a competitive bidding process, to all contractors/suppliers/entities that meet the requirements and might be interested, with or without a preselection/qualification phase.

When a public entity is contracting or when public funds are involved, some special procedures normally apply, namely a competitive open bidding (or limited invitations depending on the estimated cost of the contract), in order to ensure equality of treatment between economic operators, and transparency of processes. In this case, 'public entity' stands for the state, regional or local authorities, agencies or bodies governed by public law, or associations formed by one or several of such authorities. For that reason, EU countries are obliged to implement guarantees in their national legal systems, through the transposition of specific Directives of the European Parliament and the Council. Particularly relevant is Directive 2004/18/EC on the coordination of procedures for the award of public works contracts, public supply contracts and public service contracts, which states that the criteria to award public contracts must be the lowest price or the most economically advantageous tender.

2.3   Type of documents

Once the decision to contract construction services is taken, the client prepares all the contracting documents that, in the case of a traditional procurement, should clarify:

- the complete description of the type of work to be performed, with the aid of drawings, layouts, models, calculations, etc.
- the quality of work to be performed, through specifications and the performance required of materials, or the construction components, as well as the standards of the work expected
- the estimated cost of the work to be done, which can be achieved through the use of standard forms of measurement, percentage rates and the bill of quantities of the work to be done
- the construction programme (if applicable) and the proposed schedule, which defines the duration and the sequence expected for each part of the construction work on site
- the general and particular conditions of the contract, which clarifies the rights and obligations of the parties, as well as the risk allocation between them, in respect to the cost, time, quality, health and safety, and also the conditions imposed to suppliers, subcontractors, insurances, guarantees, price fluctuations, access to site, change and variation orders, claims resolution, etc.
- the formal bidding/tender documents: invitation for bidders, instructions to competitors, tender form or letter of tender, agreement form, etc.
- other documents that assist bidders to better understand what is expected of the job to be done, including the most relevant temporary works and other contractual obligations (health and safety procedures, environmental plan, archaeological prospection, etc.).

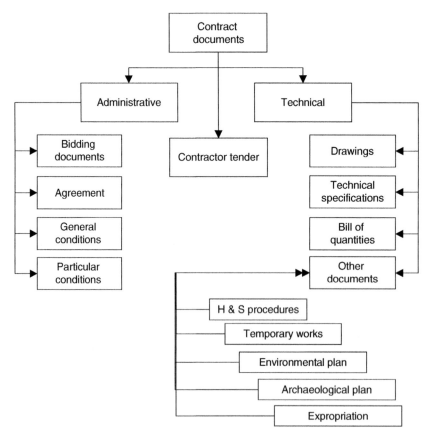

Figure 2.1   Contract documents.

This set of documents can be divided into two main groups: the administrative procedures and the technical documents (Jackson, 2004). Figure 2.1 summarises the type of documents that are usually included in a traditional construction contract.

In traditional procurement systems, these documents are usually prepared by the design team on behalf of the client. However, in respect to major clients, especially public entities, it is common to use general and particular conditions, as well as tender documents, adapted from industry standard codes to fit their own needs (Box 2.1, adapted from www.estradasdeportugal.pt). The procedure is different in other type of procurement systems (e.g. design-build), where the specifications of work to be performed are not integrally decided at the time of the bidding and the client just has a master programme to follow: in this case, competitors must present some of the documents, and the proposed solutions are used to evaluate and compare the tenders.

In more complex contracts that can include activities besides construction (such as design, finance, maintenance or operation of the facility), contractual documents are more wide-ranging and involve not just administrative and technical issues, but also policy, economic, financial and legal aspects.

> **Box 2.1   Particular conditions for road works**
>
> The public Portuguese concessionaire for the construction, repair, maintenance and operation of the national roads has developed particular conditions applicable to construction works based on standard forms, in order to complete the mandatory general conditions. They aim to regulate the following aspects:
>
> - name of the project, contract price and time for completion demanded for the works to be performed
> - mandatory insurances
> - advance payment and reimbursements
> - errors and omissions, changes, variations and adjustments
> - schedule, works programme and payments cash-flow
> - the client and the project manager
> - the construction manager and the contractor team
> - mandatory records regarding the evolution of the works
> - measurement, payments and adjustments in construction costs
> - mandatory load tests
> - maintenance during the construction period and temporary road signalling
> - debris removal and environmental site restoration
> - drawings 'as built', final specifications and materials/equipment guarantees
> - quality control, testing, inspection and laboratory materials
> - land, site offices or installations provided by the client
> - complete description of the works to be performed and contractual documents.

For instance, the invitation to present a tender or to bid in an open public procedure of a PPP contract should include at least the following documents (Yescombe, 2007): general legislative and policy background; project justification (what is the perception of a public need better fulfilled by a private partner); future service requirements; eventual support provided by public partner (financial, if the project is not viable by itself, or of another kind, such as the construction of an infrastructure); draft of the PPP contract, including risk transfer provisions, pricing formula, incentives and penalties; performance specifications regarding construction, maintenance and operating; overall project schedule and milestones; demand forecasts; provisional business plan; programme for site visits, bid meetings and procedure for clarifications; form of the bid, requirements to participate, bid deadline and bid evaluation criteria.

Also each competitor that wants to present a tender should clarify in a separate section: the detailed design and technology to be used; the detailed construction programme; the service standards and delivery; maintenance and operating manual qualifications of the main consortia, shareholders, investors, subcontractors, vendors and suppliers, as well as the management structures for all phases of the project; quality and safety assurance procedures; commercial viability of the project (based on expected demand); project costs, financing strategy and structure (internal rate of return, net present value); insurance coverage; and service fees expected.

All of the preceding relates to the contractual relationships for the provision and construction of works. However, as noticed in the precedent section, a very important part of any construction project is the provision of services (the design team, the project manager or different consultants), which must also be formalised through a contract. Apart from professional associations (engineers, architects, quantity surveyors) that have their own standard forms of agreement or terms of appointment, which can be invoked whenever one of its members gets a contract, for consultants involved in a construction project the contractual documents should clarify what tasks they are expected to do. These can include: outline and overall management of the design; budget cost and project appraisal; scheduling and planning; bill of quantities preparation; establishing the list of selected bidders; preparing the tender documents; bid evaluation and advice on the award of the contract; overall coordination of construction contract monitoring; schedule analysis and quality inspection of work on site; measurement of works and certifying for payments; and arbitration in disputes (adapted from Woodward, 1997).

Since the construction industry makes extensive use of contractual documentation, it is important to use simple language, instead of more legal terminology. Expressions like 'could', 'may' or 'reasonable' have to be avoided because they can even affect the enforceability of the contract clause in case of disputes related to contradictions or ambiguities between documents where semantics is crucial.

2.4   Bidding documents

Once documents are prepared, the owner makes the invitations to engineering and construction firms to submit their tenders for doing the work, indicating where, when and how the bids should be presented. In the case of public projects, the rules are somehow different and, depending on the estimated cost of the works, tenders are only acceptable through open competitive bidding, meaning that some rules have to be followed concerning publicity and allowance of bidders.

Because of the commitment of the owner to get the facility constructed within a given budget, before an established deadline and ensuring the maximum quality level, the notice to bidders is only sent to those contractors who are capable of doing the job, accordingly to the records of the owner (financial capability, previous jobs, experience, etc.). In the case of public projects, the intention to contract must be revealed, either in official journals, a well-known newspaper or specialised magazines addressed to the contracting community; the owner should allow whoever wants to present a bid, given that the competitor respects mandatory conditions, and technical and financial capabilities.

The eligibility to present a public tender is normally related to business and market permits. For instance competitors should have a clean criminal record, no debts to the tax authority or to social security, and must provide a valid registration licence to work as a contractor. For some public clients, they should also declare that they are not insolvent, do not employ an illegal workforce and haven't been condemned for improper professional conduct. Moreover, to evidence financial capability, audited annual accounts should be available, in order to identify annual revenues, accumulated debt, equity and economic ratios related to the operational results. Regarding technical capability, competitors are deemed to prove experience in identical works of similar cost during recent years, and provide the CVs of the main technical staff, among other relevant issues.

For those contractors who are eligible to compete, they must respond to what is asked by the client in the bidding/tender documents, which consist of the written information regarding the bidding process. These documents are part of the administrative contract documents and should include (Brown, 2003):

a. Invitation for bidders: a letter, notice or announcement to interested parties or placed in public locations, regarding the project and the requirements to present a bid. Box 2.2 (adapted from Thomas, 2001) specifies the data to include in the invitation.

Box 2.2   Content of invitation for bidders

- name of the project
- name and address of the client or the responsible for the tender analysis
- general description of the work to be performed, materials to be supplied or services
- type of procedure (open tender/close tender/competitive tender/ direct tender)
- works location
- estimated cost of the work, and the period for completion
- date, time and place where tenders should be submitted and will be opened
- how and when tender documents can be obtained, and their cost
- tender evaluation criteria and tender expiration date
- mandatory insurance and warranties
- legal form of appeal against an adverse decision regarding the procedure.

b. Instructions to competitors, meaning that the requests that should be attended to, regarding bid process, that should clarify the procedures for submitting questions and obtain clarifications about contract documents, eventual addenda to those documents provided before the tenders are open, rules concerning tender submission and rejection, notice of the winner, documents comprising the tender and so on. A complete instruction procedure for competitors is explained in Box 2.3 (adapted from Brown, 2003).

c. Tender form or letter of tender: this is the formal way (a letter of intention) whereby competitor indicates the bid price to execute the works, the completion date, the acceptance of the conditions within the contract documents (design, specifications, technical performance), the nominated subcontractors or any other requirements demanded by the owner.

d. Agreement form: the document (in a form of a contract) that will actually be signed by both parties (the contractor who has won the competition and the client) after the bidding process is concluded, which is normally an industry standard form, depending on the type of contractual arrangement and the procurement procedure.

Box 2.3   Complete instruction procedure for competitors

- complete description of the work to be performed
- to whom to submit questions, their form and the deadline, and where to obtain clarifications about contract documents, including errors and omissions of bill of quantities
- documents that prove the eligibility of the competitor to bid
- scope, content and format required of the tender
- desirable requirements of bids or bidders which would be advantageous but not critical to eligibility (e.g. special skills, experience)
- detailed criteria and sub-criteria of tenders evaluation
- documents comprising the tender (schedules, cash-flow projections, construction processes)
- admission (or non-admission) of alternatives to the client construction solutions
- procedures for managing late submissions
- definitions of non-conforming bids and procedures for managing them
- procedures for formal and informal contact between client, project team and competitors, including the notice of the winner
- other procedures regarding the provision of information to bidders, including confidentiality
- the right to terminate the tender by client, and its consequences.

2.5   Contractor tender or bid

After the analysis of the documentation prepared by the owner, the decision whether or not to present a bid must be taken. This is a major executive and financial decision that should be carefully evaluated since it implies incurring substantial costs associated to tender preparation, which may not be recovered. Normally, presenting a bid requires commitment of many man-hours by the contractor's staff in obtaining the bid documents prepared by the client, site investigation, reviewing drawings and specifications (or producing them, depending on the type of contract) and estimating the bid price. This cost, time and effort will only be recovered if the work is awarded, so the accuracy of estimating and careful bid preparation is critical (Wallwork, 1999).

In spite of the great body of knowledge on the theory of sealed bidding competitive auctions (Harris, 2006), either from the winning probability point of view (by fitting probability density functions to the bids) or economic decision theories (game theory assumptions or Monte-Carlo simulations), the competitor is never certain about the results (Skitmore, 2004). The variables of the process are the estimated cost, desired mark-up, expected profit, number of competitors and past bidding prices. For most of the construction companies the decision to bid is taken considering:

- goals and actual capabilities of the company (type of work, plans for growth, market conditions and expected return)
- physical location of the project (if it is close to the office, there is a better knowledge of the labour market, general conditions and access to the site)
- time, place and cost of presenting the bid
- method and cost of obtaining the contract documents and specifications
- legal and other official requirements necessary to present the bid
- scope of the construction works (global size, major units of work or resources needed, risk, visibility/impact of the project, etc.)
- who is the owner (regular client, payments on schedule, reputation, plan of investments, etc.) and the architect/design team (well known, past works, etc.).

If the decision to present a tender is taken, then the contract documents should be carefully reviewed, particularly drawings and bills of quantities (if they exist, according to the procurement procedure) in order to get an accurate estimate of the cost and the time needed to complete the project, since it will be the basis for determining the bid price (Kerzner, 2009). The next step for the competitor's organisation is the selection of the project manager who will guide all the tendering and the estimating process.

2.6   Estimating process

The estimate is a statement, based on the best information available, of the probable quantities (if there is not a bill of quantities) as well as the cost of materials, labour, equipment, subcontractors, overheads, taxes and profit, necessary to complete a construction project, according to the contract documents (Carr, 1989). In the construction process, estimating is an important task from the competitor's point of view. In fact, the real cost of the work will only be accurately known after the construction is completed, and the contractor will only achieve the expected profit if the estimated cost turns out to be less than or equal to the actual cost of the project.

Whenever the design is to be prepared by the construction company, and the procurement method is traditional (design by an architect on the behalf of the client, open competitive tenders, build by contractor), it is essential to achieve the correct bid price. To comply with this, the quantities of materials and work, whether developed from drawing or from bills, should be taken into account, the necessary construction methods and processes to perform each construction tasks should be analysed, fulfilment of the technical requirements should be assured, and knowledge and past experience of similar projects should also be exploited.

The assigned estimator should read carefully all the contract documents, must have the correct knowledge of construction techniques, should know the real work conditions and be familiar with construction products, materials and processes. Organisational skills and the ability to meet deadlines and work under pressure are also relevant, because normally the time to submit the tender is very tight in an environment of high competition.

Besides the total bid price, there are other cost elements resulting from the estimating process (when applicable, depending on the type of contract), which complete the tender:

- unit prices of each of the different quantities tabulated in the bill of quantities, resulting from the direct costs of performing the work
- provisional sums for eventual unforeseen work outside those items, for instance costs of a day work for different classes of labour force (carpenters, masons, painters) and equipment, or for specific material supply
- percentage for site and home office overheads
- venture risks and opportunity costs (if any)
- expected profit.

The **direct costs** are those that can be imputed to a specific work activity and therefore are not incurred if the activity is not performed. Usually they can be divided into labour, materials and equipment, as well as subcontractor costs (in respect to the main contractor tender) (Humphreys, 1991):

- labour costs are the total amount paid to the field personnel (carpenters, labourers, painters, masons) who perform the work, and include the basic wage, fringe benefits and legal charges such as taxes or insurances
- material costs include the prices of materials incorporated in the project, as well as the delivery charges. Usually the prices are given by suppliers or vendors as a specific amount within a given period and a specified location

- equipment costs includes the lease costs and fuel, if the equipment is rented or leased. If it is owned by the contractor, it includes investment depreciation and insurance, as well as operating costs (fuel, repairs and maintenance)
- subcontractor costs are the costs provided by other companies to perform a specific portion (or the total) of the work, that the contractor is not able to perform (or strategically is not interested in performing) with its own means.

Indirect costs consist of project-related costs that cannot be assigned to a specific work activity, and those expenditures of the construction company that would have been incurred even if the work had not been performed. The former are called the site overheads which can be economically traceable to a determined project and, normally, would not occur if the construction project had not been executed; the latter are the general overheads that aim to cover company's home office costs, which occur independently of each project and include all administrative issues (Harris et al., 2006):

- site overheads: site staff including transportation, cleaning site and rubbish, tower cranes and other equipment not included in work items, scaffolds, small tools, site accommodation and utilities, temporary works, welfare, first-aid and safety provisions, final clearance and handover, project risks (abnormal weather, cost increase, low productivity), and bond costs
- general overheads: general insurances, taxes, other legal costs, home office rent or lease, utilities, communications, donations, accounting expenses, advertising and marketing, salaries and other costs of the home office staff and administration board.

The estimating process ends at this phase, as the next allowances to get the bid price, namely venture risks or contingencies and opportunity costs, as well as the profit percentage, are project management and administration board tasks. Venture risks or contingency are risks related to the possibility of making a loss on the project due to unforeseen conditions, and their consideration essentially depends on the general evaluation of the robustness of estimating, professional practice and experience. Opportunity costs are those costs due to alternative choices related to the other projects of the company, and can be incurred if the competitor chooses to bid (and wins) on one specific construction project and not on another one. Profit is the possible percentage that the project will give to shareholders of the company for their investment in the construction business, and it must be determined by the administration board, taking into account the actual market conditions.

While bid price is determined after a thorough estimating process, considering that the achieved price is unique and the best choice for the particular circumstances of the company and the evaluation made of project costs, risks and expected profit, the project manager prepares the necessary documents

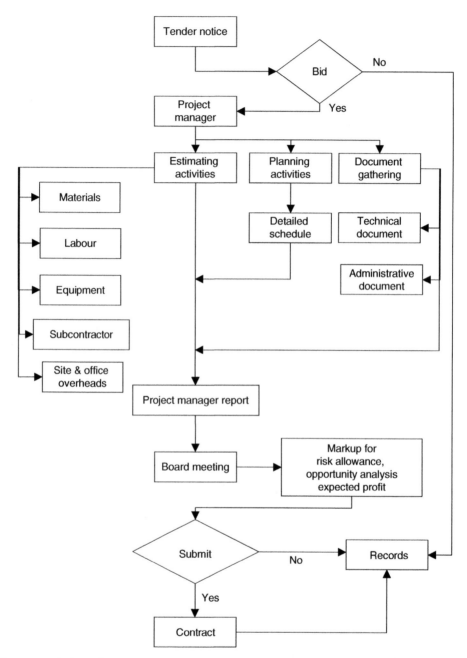

Figure 2.2 Schematic procedure of the tender process (adapted from Harris et al., 2006, and Thomas, 2001).

to complete the tender. (Figure 2.2 shows a schematic procedure for the tender process in traditional construction procurement.)

Besides estimating activities and gathering together all types of documents requested in the procurement process (health and safety plans, quality assurance control, environmental programme, capability documents, master plans or design sets, financial agreements, etc.), the competitor should also present a provisional schedule, with the proposed sequence for the main tasks of the construction works, as well as the total duration expected of the project.

In the first stage, planning concerns the identification of all the activities needed to build the facility, the strategic approach and time and effort needed to complete the job, while scheduling is related to the sequencing of those activities, assigning expected durations and relationships between them, as well as the minimum amount of resources needed to complete each task in the most efficient way (Antill and Woodhead, 1982). During the tendering procedure, the objective of the project manager should be to present a preliminary schedule, with sufficient detail to allow the client to evaluate the operational options and compare them with other competitors' schedules.

After a final review of the administrative and technical documents, schedule, job quantities, unit prices and the achieved bid price, the final step is the tender submission, while the project manager and the administration board try to anticipate what other competitors will do, and take the last-minute decisions in relation to risk assessment and mark-up assigned to the project. A possible strategy could also include the presentation of an unbalanced bid (Cattell et al., 2008) which refers to raising unit prices on items to be completed in the early stage of the project and lowering the unit prices on items to be completed in the later stages, in order to ease the construction financing. At the same time, if competitor feels that some item of the bill of quantities is underestimated in amount, so that a high unit price on that item would increase profits, an unbalanced bid may also occur.

In more complex projects, the preparation of tender documents and the stipulation of the final bid price need extra effort, because it is common for the main contractor to need to associate with either another contractor(s) (to take advantage of their permissions and licences) or a design team, in the case of design-build procurement, or even with operating companies or financing entities, in some types of PPP projects. This type of partnering involves complex negotiations, the production of several types of documents (drawings, technical proposals, financial closeouts, measurements, definition of a business plan, stipulation of consortia agreements, long-term investment cash flows, etc.), in a very short time, as tender schedule is normally very tight.

2.7 Contract agreement

The basis of the construction contract is the agreement, which is composed of two parts: the offer (bid or tender form) and the acceptance. The offer is more than an attempt to negotiate, since if it is accepted, it will become binding. The acceptance is normally assured through a letter from the client issuing the notice to proceed. However, to be effective, the contract should be formalised by the signing of the articles of the agreement. In a legal sense, the formal contract agreement is the single document that binds the parties, client and contractor (Ashworth, 2012), and should include the following:

1. names of client and contractor (possibly architect/engineer/quantity surveyor)
2. date of signing the contract
3. location and nature of work to be executed
4. amount of the contract sum
5. complete list of contract documents relevant to that particular construction contract, namely:
   a. biding documents
   b. tender of the winning competitor
   c. general conditions
   d. particular conditions
   e. technical specifications
   f. bill of quantities
   g. construction drawings
   h. other technical or administrative documents.

Similarly to other components of contract documents, there are standard forms, prepared by either industry partners or by major clients (especially public ones) for the clauses of agreement, applicable to different types of potential construction contracts (Box 2.4 shows an example from FIDIC, 1999). If the parties make any amendments to these clauses of agreement or to any other part of the contract, then the adjustments should be signed by both parties.

Box 2.4   Contract Agreement (reproduced with permission from FIDIC)

This Agreement was made in_____, ___ (day) of _____ of 20____ Between_____ (name) of _____ (hereinafter called 'the Employer') of one part, and_____ (name) of _____ (hereinafter called 'the Contractor') of the other part

Whereas the Employer desires that the Works known as_____ should be executed by the Contractor, and has accepted a Tender by the Contractor for the execution and completion of these Works and the remedying of any defects therein,

The Employer and the Contractor agree as follows

1. In this Agreement words and expressions shall have the same meanings as are respectively assigned to them in the Conditions of Contract hereinafter referred to.

(Continued)

2. The following documents shall be deemed to form and be read and construed as part of this Agreement: the Letter of Acceptance dated_____, the Letter of Tender dated _____ the Addend(s) identified by _____, the Conditions of Contract, the Specification, the Drawings, the completed Schedules
3. In consideration of the payments to be made by the Employer to the Contractor as hereinafter mentioned, the Contractor hereby covenants with the Employer to execute and complete the Works and remedy any defects therein, in conformity with the provisions of the Contract
4. The Employer hereby covenants to pay the Contractor, in consideration of the execution and completion of the Works and the remedying of defects therein, the Contract Price at the times and in the manner prescribed by the Contract

In Witness whereof the parties hereto have caused this Agreement to be executed the day and year first before written in accordance with their respective laws

SIGNED by:_____   **SIGNED by**:_____
for and on behalf of the Employer for and on behalf of the Contractor

WITNESS_____   **WITNESS**_____
Name: _____   Name: _____
Address: _____   Address_____

2.8 Bill of quantities

Bills of quantities are documents, prepared by the design team, that assume the form of codified spreadsheets, comprising a list of items of work to be carried out. They provide a brief description of the type of the work, the quantities of the finished work and the units of measurement (Law, 1994). Thus, the objective of this document is:

- to provide sufficient information on the quantities of works to be performed to enable bids to be prepared efficiently and accurately
- to allow each contractor who is bidding for a project to price construction work based on exactly the same information as other contractors using a minimum of effort
- to provide unit prices and consequently a priced bill of quantities for use in the periodic valuation of the works being executed.

In order to achieve these objectives, works should be itemised in sufficient detail allowing for distinguishing between different classes of work, material or construction processes, as well as between works of the same nature but carried out in different locations or in other circumstances which may give rise to different considerations of cost. Any work that is part of a civil engineering contract should be measured in accordance with the appropriate method of measurement for the particular tasks involved in the completion of the work. So the correspondent item should be included in the bill of quantities, with a description stating the units, the criteria and how it can be measured (Murdoch and Hughes, 2008).

The layout and content of the spreadsheet should be as simple and brief as possible, and quantities should be computed from the drawings, rounded up or down as appropriate; spurious accuracy should be avoided, as they are intended to be estimated and provisional quantities to provide a common basis for bidding. To present the offer during the bidding process, the contractor should provide a rate or a price against each item in the bill of quantities, whether quantities are stated or not, which must include all plant, labour, supervision, materials, equipment, maintenance, insurance, profit, taxes and duties, together with all general risks, liabilities and obligations set out or implied in the general or particular conditions of the contract. A typical form of a bill of quantities is presented in the Figure 2.3.

Frequently, in addition to the main table with codification, description of the work, units, quantities, rate and total amounts, bills of quantities have an introductory explanation about how they should be filled in, and the bases that were used to measure the work. For instance, descriptions ought to identify correctly the work covered by the respective items, but the exact nature and extent of the work has to be ascertained from the drawings, specifications and other conditions of the contract, which may be read in conjunction with the work classification.

Furthermore, a detailed description of an item may be omitted, if a reference is given in its place which precisely identifies where the omitted information may be found on a drawing or in the technical specification. Also, when the semantic description is not enough to precisely define an item of work to be performed, additional description may be given to identify the specific work by reference to its location in the drawings or in some other contractual document. Additional description may be provided, where the item is outside normal and sequential codification, related to the nature, location, importance or any other special characteristic of the work that is likely to

Number	Item description	Unit	Quantity	Rate	Amount (partial)	Amount (total)

Figure 2.3   Typical form of a bill of quantities.

give rise to special methods of construction or considerations of cost, not normally considered in the items list.

Another specification frequently contained in the bill of quantities, is that related to the units applicable to each item (m, t, kg, m², number …), and whether quantities and unit prices (or rates) should be expressed in whole numbers or with decimal places. For construction facilities or linear projects such as highways, utilities installation or tall buildings, where quantities of repeated works can easily reach millions (earthworks, steel, pipes), this aspect can be decisive to winning or losing the contract.

Because bills of quantities generally use industry standard forms, they may include several items and types of works that don't apply to the particular tender procedure under quotation. Also, some of those standards include not just the description of the finished works or the finalised construction process, but items that are generally assigned as indirect costs or temporary works. Therefore, the competitor should carefully place their prices (or rates), beware of any missing or duplicated pages or items, because no liability whatsoever will be accepted by the client in respect to errors in a tender due to such issues. Likewise, on no account can the client (or designer) be held responsible if these documents were to be used for placing orders for materials or subcontractors; this is solely at contractor's own risk, as quantities placed by client are just estimates.

The bill of quantities is only really necessary for the unit-price type of contract, because in other contracting systems (e.g. cost plus fee or lump sum) the construction works can be priced in other ways. Thus, it is not mandatory that all clients/designers prepare bills of quantities, which in fact depends not only on the type of contract but also on the project coalition (i.e. in the design-build contracting the detailed bill of quantities is prepared by each construction company). Bills of quantities can also be used for other purposes, as diverse as the preparation of interim valuation of works, cost control, valuation of change orders, ordering of materials from vendors, obtaining subcontractors quotations or quantification of contract claims.

The generalised use of bills of quantities to price constructions works led to the development of industry standard items for different types of construction work, materials and processes, sometimes associated with the technical specifications. Normally, this work is done in each country by the national standardisation bodies, construction industry and professional associations, research institutions and major construction (public) owners.

2.9   General and particular conditions

The general conditions section of a contract for construction summarise the rights and duties of the parties, delineate responsibilities and clarify the specific terms used in the other contract documents. In fact, certain stipulations regarding how a construction contract should be administered and the relationships between the parties are often the same for all contracts. For that reason, industry and professional organisations have developed a set of standard general conditions to be used either in private or public contracts (Ashworth, 2012). It is also common for contracting bodies that usually award several construction contracts per year, to develop standard sets of contract conditions, distributing them to their stakeholders. The same situation occurs with public entities, being common in some countries to use mandatory general conditions for public contracting of construction services, previously approved by law or regulations.

The most used general conditions are well known to contractors and owners that works in a particular country or region, and because past disagreements or disputes over the interpretation of clauses have probably been decided by courts or arbitration, it is preferable to use one of the forms available (see FIDIC, 1999), rather than producing a personalised and specific one. Even if the client, or the uniqueness of the construction project, points to some different provisions, this can be made by adding special clauses to the general conditions adopted. For that reason, it is also recommended not to impose new clauses that are biased in favour of the client, as contractors will tend to overprice their bid, in order to overcome the additional risks involved.

Although these forms are often very similar, they are different in details. Indeed, the specificity of the type of construction works (civil engineering, mechanical, electrical, building, etc.), the dimension of the project (small, medium or large) or the procurement method used (design-bid-build, design-build, etc.), has made it necessary to develop general conditions of contracts to better fit each situation. However, international reports have recently highlighted some disadvantages of the use of a wide variety of standard forms, because of the duplication of effort, wasteful use of resources and the necessity of using expert or judicial decisions to interpret some of the clauses.

There are some standards internationally adopted, especially in English language countries (Thomas 2001, Ashworth 2012): FIDIC (International Federation of Consulting Engineers), JCT (Joint Contract Tribunal), ICE (Institution of Civil Engineers), NEC/ECC (New Engineering Contract), AIA (American Institute of Architects), AGC (Association of General Contractors), EJCDC (Engineers Joint Contract Documents Committee), World Bank, United Nations, the Government Prime Contracting time and so on. For instance, the well-known FIDIC General Conditions of Contract for Works Designed by the Employer (FIDIC, 1999) are divided into 20 main sections, each of them with several clauses as described in Box 2.5.

The examples in Box 2.6 (adapted from Hendrickson, 2008) illustrate typical contractual language, where opposite risk allocation is considered between the contractor and the owner. In the first one, the contractor is exposed to a high risk level, since they are responsible for every occurrence causing damage to the owner (except for negligence), with no limited amount, and independently of insurance protection, as well as for different site conditions involved during the works. However, in the bottom clause the contractor's liability is limited to whatever can be collected from the insurance company, and the unit prices for foundations and earthworks should include prospecting and investigations made by the owner.

Box 2.5   Main items considered by the FIDIC General Conditions of Contract for Works Designed by the Employer (reproduced with Permission from FIDIC)

- general provisions: providing fundamental definitions about the contract, applicable laws, interpretation, priority of documents, communications, language, supply of documents, confidentiality, joint liability, etc.
- client responsibilities: information and services that client is required to supply, like access to site, permits, licences or approvals, personnel, financial arrangements and owner's claims
- engineer responsibilities: duties, authority, delegation, instructions, replacement and determinations of the owner's representative
- contractor responsibilities: general obligations regarding construction procedures and supervision of the work, labour, equipment and materials, subcontractors, setting-out, safety, quality assurance, environmental protection, site data and different site conditions, transport of goods, cooperation and avoidance of interference, progress reports, access routes, fossils, etc.
- subcontractors: nominated subcontractors, payment and evidences
- staff and labour: rates of wages, labour conditions, labour laws, working hours, facilities for staff and labour, health and safety, contractor's superintendence and personnel, records, disorderly conduct, etc.
- plant, materials and workmanship: manner of execution, samples, inspection, testing, rejection, remedial work, ownership of plants and materials, royalties, etc.
- time requirements: notice for commencement, time for completion, schedules, delay damages, extension of time, suspension of works, resumption of works, etc.
- tests on completion: contractor's obligations, delayed tests, re-testing, consequences of failure to pass the tests on completion
- taking over: necessary proceedings for taking over the completed works by the owner
- defects liability: dealing with contractors' obligations to remedy defects
- measurement and evaluation: defining the works to be measured and the methods, as well as the procedures for pricing and evaluation quantities of work
- variations and adjustments: the right to vary and issue change orders, value engineering procedures, evaluation of variations, adjustments in legislation and changes in costs
- payments: contract price, payment period, advance payment, interim certificates, payments scheduling, retentions, final payments, discharges, etc.

(Continued)

- termination by client: regulates the entitlement to terminate the contract, for convenience or through the fault of the contractor
- suspension and termination by the contractor: discusses the situations that give contractor the right to suspend or terminate the work
- risk and responsibility: defining what client's and contractor's risks and liabilities are, their consequences and limitations, including those relating to intellectual and industrial property rights
- insurance: some of the risks and liabilities must be insured, such as the works, equipment, injury to persons or property and working staff and personnel
- force majeure: an exceptional circumstance beyond the control of a party in the contract which cannot be reasonably expected and avoided
- claims, disputes and arbitration: the mandatory proceedings that parties should use to attempt to resolve any claim arising from the construction contract.

Box 2.6   Example of high and low contractor risk

High contractor risk:
Except when the negligence of the OWNER is involved or alleged, the CONTRACTOR shall indemnify and hold harmless the OWNER, its officers, agents and employees, from and against all loss, damage and liability, as well as from any claims for damages on account of bodily injury, including death. The protection should be extended to all damages to property, including property of the OWNER and third parties, direct and/or consequential, caused by or arising out of, in whole or in part, or claimed to have been caused by or to have arisen out of, in whole or in part, an act of omission of the CONTRACTOR or its agents, employees, vendors, or subcontractors, of their employees or agents, in connection with the performance of the works, whether or not insured against. At the same time, the CONTRACTOR, at its own cost and expense, shall defend any claim, suit, action or proceeding, whether groundless or not, commenced against the OWNER by reason thereof or in connection therewith.

Subsurface information such as boring logs, tests and results of other investigations are available for inspection, but whether included in the plans, specifications or otherwise available to bidders, was obtained and is intended solely for owner's design and estimating purposes, and each bidder is solely responsible for all assumptions, deductions or conclusions made from this information.

(*Continued*)

> *Low contractor risk:*
> The CONTRACTOR hereby agrees to indemnify and hold the OWNER and/or any parent, subsidiary, affiliate or the OWNER and/or officers, agents or employees of any of them, harmless from and against any loss or liability arising directly or indirectly out of any claim or cause of action for loss or damage to property including, but not limited to, the CONTRACTOR's property and the OWNER's property and for injuries to or death of persons including but not limited to the CONTRACTOR's employees, caused by or resulting from the performance of the work by the CONTRACTOR, its employees, agents and subcontractors and shall defend the OWNER in any litigation involving the same regardless of whether such work is performed by the CONTRACTOR, its employees or by its subcontractors, their employees, or all or either of them. In all instances, the CONTRACTOR's indemnity to the OWNER shall be limited to the proceeds of the CONTRACTOR's umbrella liability insurance coverage.
>
> Subsurface information such as boring logs, tests and results of other investigations are available for inspection, but whether included in the plans, specifications or otherwise available to bidders, however obtained for owner's design and estimating purposes, should be considered in the assumptions to price the foundations and earthworks.

Those aspects of the contract relationship that are peculiar, unique or specific to a project are described in the particular, special or supplementary conditions of the contract. This includes items related to project duration, additional instructions related to the commencement of works, mandatory wage rates for a local area, client-furnished materials, facilities or services, formats required for progress report and schedules, soil testing and information, traffic control and pedestrian safety requirements, phasing or special scheduling needs, penalties or liquidated damages and so on. Some of these particular conditions are extensions or interpretations of the general conditions, and the clauses are sometimes grouped under the same title, resulting from written changes in the general conditions, or even inserted as additional subclauses. For instance, FIDIC (1999), with the published Conditions of Contracts for Construction, also presents the guide for the preparation of particular conditions, which proposes the deletions, changes or additions to the clauses of the general conditions, according to the specificity of the project.

2.10   Technical specifications

Whereas the general conditions, particular conditions or bidding documents deal primarily with the administrative aspects of the contract, the technical specifications refer directly to the construction itself. In fact, they should be used in conjunction with the drawings, and their intent is to describe in detail the technical aspects of the work to be performed. So the technical specifications should identify the requirements of the construction project, defining technical characteristics required in respect to labour, materials, equipment and construction processes, aiming to clarify and describe the following aspects: quality of materials, standard of workmanship, methods of installation and erection, quality control and quality assurance procedures (Ashworth, 2012).

To ensure international quality standards, each component of the final product (the built facility) must be identified and specified (Hinze, 2010). Often, the quality requirements of construction materials or products are established by reference to accepted industry practices, national technical specifications or to common international standards. For materials, components and equipment integrated in private projects, this can be done by citing a specific brand name or model number. In public procurement, this practice is not allowed and it is mandatory to characterise materials or other products either by reference to technical specifications relating to the design, calculation and execution of the works and use of the products, followed by the words 'or equivalent', or in terms of performance (physical and mechanical demands) or functional requirements of the construction products, in order to protect open competition and equality of opportunity among bidders. Furthermore, European Directives state that, unless justified by the specificity of the contract, technical specifications shall not refer to a specific brand or source, or a particular process, or to trademarks, patents, types or a specific origin or production with the effect of favouring or eliminating certain undertakings or certain products. Such reference is only permitted on an exceptional basis, where a sufficiently precise and intelligible description of the subject-matter of the contract is not possible; once again, such reference must be accompanied by the words 'or equivalent'.

In respect to construction processes, basic constitution of materials, their components, physical and mechanical behaviour, testing and samples, each country's standardisation bodies generally develop tools to assist industry, and particularly designers and constructors, either for building or civil engineering. Through these tools, all works are organised and managed in written standard formats, which helps not only to organise and communicate design data information, but also to track project costs, and above all to guarantee quality standards. It is also common that under each work type or product specification, a system of numbering is implemented (Jackson 2004), ranging from a broad scope to a more narrow one, that allows industry practitioners to have knowledge and experience of the descriptions of materials, products and work procedures, included in the technical specifications.

For example, the types of work, construction processes or materials normally used in buildings are: general requirements, site work, concrete (Box 2.7), masonry, metals, wood and plastic, thermal and moisture protection, doors and windows, finishings, equipment, furnishings, special construction, conveying systems, and mechanical and electrical works. For usual

> **Box 2.7   Techical specification for concrete works**
>
> In spite of the recent increasing use of other construction material (steel, wood, glass, etc.), concrete is still the most used construction material worldwide. The usual technical specifications regarding production, manufacture and placing of concrete, should detail at least the following aspects:
>
> - characteristics of materials: cement, aggregate, water, admixtures
> - proportioning and manufacture of concrete
> - time between mixing and placing
> - transport and placing
> - compaction
> - field tests, curing and joints
> - preparation of surfaces prior to the placement of concrete
> - construction tolerances and forms
> - cold and warm weather concrete
> - repair of new concrete
> - surface finish
> - quality control, testing, inspection and laboratory equipment.

transportation infrastructures such as roads and highways the technical specifications usually refer to the following type of civil engineering operations: earthworks, drainage, pavements, bridges and viaducts, crossovers, signalling, contention works and tunnelling.

2.11   Contract drawings

The drawings are graphical or schematic indications of the construction work to be performed in the stipulated contract (Murdoch and Hughes, 2000). Their objective is to communicate the insights of the designers (architects and engineers) to the contractor. They are the single most important communication tool employed by the construction industry to define and specify the details of the construction process. In the USA, drawings are also called plans or blueprints (Hinze, 2010).

In the traditional contracting system, the drawings should ideally be finalised before the procurement stage. However, this is not always the case, due to insufficient time available for pre-contract design, or because of the indecision of the client or the design team, about the particular solutions. In these situations, some preliminary drawings are presented to the bidders, with the most accurate information to enable them to offer the price for the works, and then, after the contract has been awarded, but before the beginning of the related part of the construction, the definitive drawings are presented. In other types of procurement systems (design-construction for instance) drawings, and specially the detailed ones, are only completed by the time they are needed; the advantage of these systems is the control by the contractor of both the construction and the design phase, because he can easily anticipate the needs of the design details, in order to guarantee that the work can flow without disruption.

To ensure that information presented in the construction drawings is quickly and easily understood by its main users, contractors, subcontractors and vendors, there are some international standards about the preparation of these documents. In fact, standardisation eases their use by everyone involved in construction, and helps maintain a consistent quality of graphical communication. In EU countries, the European Committee for Standardisation developed European Standards for the size, scale and organisation of construction drawings that should be used in each country. More widespread is International Organisation for Standardisation (ISO), which has published standards for construction and technical drawings (ICS 01.100.30). It is also common for major construction clients, and specifically the public ones, to develop their own standards and methods of producing and presenting drawings, which at least should include:

- name and address of the consultant (architect or engineer)
- drawing number (for reference and recording purpose)
- scale or scales
- title or the scope of the work covered by the drawing.

Nowadays, the complexity and intricacies of some types of construction facilities make it almost impossible to figure out all the details in traditional two-dimensional plans. In fact, most drawings are actually produced electronically using computer aided design and drafting software (e.g. AutoCAD or ArcView). Three-dimensional and even 4D (including time evolution of construction) electronic modelling is common in most architectural/engineering firms, particularly for major buildings or other important facilities. The future evolution in this field is the generalised use of BIM (building information modelling), which is the use of computer technology to virtually model construction products before producing them in the site; it allows cataloguing of their physical and functional characteristics in order to give immediate design intention and information to clients, who may anticipate the future use of the building or infrastructure, and automatically evaluate it. Chapter 4 deals with these subjects in depth.

Normally, drawings are organised by different types of designs required to fully complete the construction, according to the standard protocols, and are numbered sequentially for easy reference (Jackson, 2004). While preparing the set of drawings, a cover sheet is placed at the top, providing general information about the project, such as the project's identification, the owner's name, the designer and even the financier. Then, taking for example a building, works associated with site such as grading, earthworks, demolitions, excavations, site utilities, streets, drainage and landscaping are depicted. Next are the architectural drawings, describing the aesthetics of the facility, size, dimensions, shape and overall appearance; they should include floor plans, exterior elevations, sections, doors, windows and all details needed. Structural drawings are made by an engineer and are aimed to identify and detail the major components of the structural frame of the building (columns, girders and beams). Also essential are the mechanical set where the plumbing drawings define water lines, sewer lines and gas lines, while the HVAC drawings show the heating, ventilation and cooling ducts and associated equipment, as well as the energy-efficiency requirements, elevators and fire and intrusion protection. The final type of drawings are the electrical set, which includes wiring, transformers and panel boxes, switches and light fixtures, as well as computers, communication and media networks.

In each set of drawings referred above, there are specific types that help the reader to get a better understanding of the overall construction. Because more and more information has to be communicated to explain how the facility ought to be built, each level of drawings becomes more specific, incorporating greater detail. So, there are typically four types of drawings: plans, which is a horizontal view of the part of a construction (foundation, floor, etc.); elevations, that shows what the construction should look from outside and from a given direction; section drawings, used to present a deeper understanding of a certain part of the design, through vertical cuts through the construction; and details, which depict portions of the construction, at higher scales, to explain finer elements of the design. Sometimes schedules are also prepared in conjunction with the drawings, normally to give special information about details like windows, doors, access holes or internal finishes.

Given the amount of information that must be covered in these documents, there are often discrepancies between the various levels of drawings (Box 2.8) which should be resolved by guidelines normally established in the general conditions. In case of conflict, typically the more detailed drawings take precedence over the more general ones.

Box 2.8   Design peer review

The design review is the critical analysis of the design elements by a third independent party, in order to avoid misunderstandings, impractical specifications, conflicts or drawing discrepancies. It has gained importance through the years, and in some countries there is an obligation on the public authority to ensure the design review of a public facility by someone other than its author (another design team), whenever the works assume relevant complexity or where innovative construction methods, techniques or materials are employed. The advantage of design peer review has been pointed out as a preventive measure to combat the lack of design quality, due to price concurrency and market decline, which has

(Continued)

consequences in the increase of construction cost overruns, insufficient durability of the facility and even an effective risk to life or property. For each component of the construction, the design review can be graded from a simple check list, to an in-depth consideration of all conceptual criteria, including calculation, measurements, specifications and drawings. For an ordinary building the design review should check out:

- loads, structural dimensioning including foundations
- durability of the building, its components and equipment
- energy consume in HVAC and lightning
- interior air quality
- security and safety (fire, intrusion, emergencies)
- environmental and sustainable construction.

2.12   Other documents

Besides information regarding materials and construction works, it is common to attach to the contract other types of information either prepared by the design team (supporting structures, soil test information or phasing schedules), provided by the owner (surveys, traffic statistics or security and sanitation requirements) or provided by the contractor (temporary works, environmental plan). In buildings and other major civil engineering works, a complete and reliable soil characterisation is indispensable to accurately estimate the costs of earthworks, ditches and foundations. For that, a complete geotechnical report, together with the results of laboratory and site tests, borings, ground water level, soil samples and inspections, should be added to the contract documents. Furthermore, in tunnelling, dams or transportation infrastructure construction, it is common to detail during the design phase the trench dismantling plan, as well as the blasting process to use in rock excavations. In the case of an existing structure planned for demolition, a dismantling plan should also be part of the contract documents.

Another important and mandatory document according to labour legislation is the health and safety study. In general, the baseline is prepared by the designer, on the behalf of the client, which must at least define the safety policy, the rules to observe on the site, a description of the project and the main risks identified. After the agreement is signed, the contractor has a duty to complete, develop and implement the study; the contractor establishes how health and safety will be managed during the construction phase, producing a health and safety plan. This is also a part of the contract documents and has to be completed according to the progression of the works on the site. Normally, the plan includes: health and safety targets for the project, risk assessments, training records, plant maintenance and inspection records, detailed work sequence, safety data information, emergency contact details, management of subcontractors and suppliers, site security/access, onsite training, social and first-aid facilities, reporting of accidents and incidents, production and approval of construction methods, site rules for visitors, fire and evacuation procedures (Jackson, 2004). Chapter 9 deals with the health and safety study and plan.

In some projects, the temporary works are as important as the definitive works, and must be previously detailed by the client, in order to guarantee its adequate implementation during the construction phase (Chapter 5). Some typical examples are: utilities maintenance during earthworks to provide services to the local population; road or railway phased detours to ensure that traffic can flow during construction; changes in urban road traffic and signalling; special temporary structures and pathways for pedestrian access and circulation; ditches, drainage and procedures for dewatering with wells or well-point systems; specific plans needed for excavations support or provisional retaining walls like planking, shoring, cofferdams and sheet pilling; provisional structures, such as formwork and shuttering needed to pour concrete, temporary foundations and piles in a bridge construction; protection of buildings or other construction near the site (such as provisional ducts or noise-proof screens). All of these plans and procedures can be part of the contractual documents.

Additionally, contractor, client and all the actors in the construction process are also subject to law and regulations regarding environmental impact assessments, urban and rural environment (respecting zoning procedures to residential, commercial or industrial facilities), sustainability (imposing limits to waste or operational energy), social protection of workers and amenity for users, and they should also observe the best-practice codes of construction activity (CIOB, 2010).

Another important contractual document, which is mandatory in some countries especially for public projects, is the handling, storage and treatment of construction and demolition waste, and the way it can be reduced, reused or recycled (Chapter 10. It involves the reprocessing of the materials into new products, such as brick or stone into aggregates or asphalt in new bituminous mixtures. All the other waste that is not capable of being reused or recycled, including hazardous and dangerous materials (fuels, chemicals, heavy metals), must be classified, stored and removed by proper means, according to the applicable laws and regulations.

Besides waste management, the environmental monitoring plan, which is a mandatory document, aims to help construction managers to coordinate and control the construction works in order to comply with environmental protection and sustainability, to implement the measures contained in the environmental impact study and to define measurable indicators to track the environmental impact caused by the construction activities on the site (Chapter 10 deals with the environmental plan in more detail). These indicators are very varied and the most common are: air quality and pollution around the site (dust, gases, atmospheric polluters); noise increment due to the works, particularly in residential areas during evening or at night; vibration caused by machinery and equipment, which should respect maximum bounds imposed by technical norms and regulations; excessive settlement, consolidation or deformation of soils and foundations, as well as slope stability; water quality control, water contamination, maintenance of watercourses and underground water velocity and flow rate; the maintenance and protection of animal and vegetable species; and all licences and authorisations required to do the construction work. At the end of the construction works, the plan should foresee the dismantlement of the jobsite facilities, and the removal or demolition of auxiliary facilities and temporary structures, including restoration and landscaping of the affected area.

Another document that is sometimes attached to the contract documents is the archaeological surveillance plan during the construction activities (Jackson, 2004). Typically, the dispositions of the plan strengthen the duty of contractor to keep at least one archaeological monitor on site while all subsurface disturbances and excavations are occurring. Following completion of the site, the monitor will conduct a final site check for any uncovered resources, and has the authority to stop the construction progress upon the finding of any archaeological discovery (fossils, remains, ceramics) until it is collected or protected. Special protection measures can also be foreseen for historic or classified monuments and places near to the construction site.

Sometimes, 3D models in real dimensions of a determined construction element or process, prototypes, software simulation to analyse the structural behaviour of the facility or part of it, materials samples or catalogues, and the quality control plan (Kerzner, 2009), are elements that can be important to define the required solution and should be part of the contract documents prepared by the client.

A final reference to the expropriation (which is the act of taking of the land without the consent of the landlord by a public authority in the exercise of its statutory powers) or land acquisition process, that should include specific drawings, land characterisation according to soil occupation and real estate valuation reports, will complete the contract documents, particularly in the case of major public projects.

References

Antill, J. and Woodhead, R. (1982). *Critical Path Methods in Construction Practice* (3rd Edition). Wiley, New York.

Ashworth, A. (2012). *Contractual Procedures in the Construction Industry* (6th Edition). Pearson, New York.

Brown, E. (2003). *California Public Works: Managing Risk & Resolving Disputes* (3rd Edition). Ernest Brown & Company, San Francisco (California).

Carr, R. (1989). Cost-estimating principles. *Journal of Construction Engineering and Management*, 115(4), 543–551.

Cattell, D., Bowen, P. and Kaka, A. (2008). A simplified unbalanced bidding model. *Construction Management and Economics*, 26(12), 1283–1290.

Chinyio, E. and Olomolaiye, P (2010). *Construction Stakeholders Management*. Willey-Blackwell, Oxford.

CIOB (2010). *Code of Practice for Project Management for Construction and Development* (4th Edition). Wiley-Blackwell, Oxford.

FIDIC (1999). *Conditions of Contract for Construction*. International Federation of Consulting Engineers, Geneva.

Harris, F., McCaffer, R. and Edum-Fotwe, F. (2006). *Modern Construction Management* (6th Edition). Blackwell, Oxford.

Hendrickson, C. (2008). *Project Management for Construction* (Electronic Version 2.2). Available 07/12/2012 at http://pmbook.ce.cmu.edu/.

Hinze, J.W. (2010). *Construction Contracts* (3rd Edition). McGraw Hill, New York.

Humphreys, K. (1991). *Jelen's Cost and Optimisation Engineering* (3rd Edition). McGraw-Hill, New York.

Jackson, B. (2004). *Construction Management – Jump Start*. Sybex, Alameda (California).

Kerzner, H. (2009). *Project Management: A Systems Approach to Planning, Scheduling, and Controlling* (10th Edition). Wiley, New York.

Law, C. (1994). Building Contractor Estimating: British Style. *Cost Engineering*, 36(6), 23–28.

Li, B. and Akintoye, A. (2003). An overview of public-private partnership. In: *Public-private Partnerships: Managing Risks and Opportunities* (Akintoye, A., Beck, M. and Hardcastle, C., eds.), 3–30, Blackwell, Oxford.

Lowe, D. and Leiringer, R (2006). *Commercial Management of Projects – Defining the Discipline* (2nd Edition). Blackwell, Oxford.

Murdoch, J. and Hughes, W. (2008). *Construction Contracts: Law and Management* (4th Edition). Spon Press, London.

Moura, H. and Teixeira, J.C. (2010). Managing stakeholders conflicts. In: *Construction Stakeholder Management* (Chinyio, E. and Olomolaiye, P., eds.), 286–316, Blackwell, Oxford.

Skitmore, R. (2004). First and second price independent values sealed bid procurement auctions: some scalar equilibrium results. *Construction Management and Economics*, 26(8), 787–803.

Thomas, R.W. (2001). *Construction Contract Claims* (2nd Edition). Palgrave, New York

Wallwork, J. (1999). Is there a right price in construction bids? *Cost Engineering*, 41(2), 41–43.

Woodward, J. (1997). *Construction Project Management – Getting it Right the First Time*. Thomas Telford, London.

Yescombe, E. (2007). *Public Private Partnerships – Principles of Policy and Finance*. Elsevier, Oxford

Further reading

Hinze, J.W. (2010). *Construction Contracts* (3rd Edition). McGraw Hill, New York.

Murdoch, J. and Hughes, W. (2008). *Construction Contracts: Law and Management* (4th Edition). Spon Press, London.

3 Procurement Approaches

3.1 Educational outcomes

The main objective of this chapter is to examine how the construction team is built up under alternative procurement approaches. Other educational outcomes are:

- acknowledge the procurement processes used by owners to select the project team
- highlight relationships among parties for each of those processes
- analyse the communication process on site and stress the importance of meetings.

3.2 Introduction to procurement

Procurement is the process of selecting a team for successfully conducting a construction project. Selecting the right team is becoming more and more difficult as construction projects increase in complexity. However, team building is a key factor for project success (Winch, 2010). Building the team is primarily a task for the project owner (or simply the owner, sometimes called the client or the employer). This may be a difficult task because even small projects now need a large number of designers, managers, consultants, contractors and suppliers. Moreover, a large set of different external entities must be considered at all times of the project development, such as local authorities, public service suppliers and so on.

There are three main ways in which owners procure construction services (Ashworth, 2008): in-house capability, appointment and competition. The

Construction Management, First Edition. Eugenio Pellicer, Víctor Yepes, José C. Teixeira, Helder P. Moura and Joaquín Catalá.
© 2014 John Wiley & Sons, Ltd. Published 2014 by John Wiley & Sons, Ltd.

in-house capability is extensively used by owners who undertake large volumes of work, such as large private developers and state agencies. The functions usually handled in-house are related to design and facility management, because there has been a shift towards outsourcing most project functions in recent years. There are a number of important advantages offered by the in-house capability option, including:

- no need for a complete contract prior to starting the project
- extensive administrative audit capability for the owner
- minimal risk of opportunistic profiteering at the owner's expense.

However, there are also important disadvantages of this option:

- possible diversion from the owner core business
- expensive to maintain in-house capability for small number of projects
- possible production inefficiencies because of lack of competition.

Appointment is often used for design and project management services. However, in many European countries, appointment for public projects is limited to a pre-established maximum contract value; above that value a tender procedure is required (European Commission, 2004). Appointment may also be used when the project requirements are so specific that only one entity can fulfil them or when the requirement to mobilise resources is so urgent that there is no time to go through a tendering process (e.g. urgent repair of a wall at risk of collapsing). The advantages of appointment are:

- restricted search of suppliers with lower costs and reduced risks
- high-trust relationships built up between the client and the supplier.

Possible disadvantages are:

- lower levels of production efficiency and effectiveness due to the lack of competition
- lack of transparency of the appointment criteria
- too-cosy relationships, leading to inadequately rigorous appointment criteria and this can degenerate into corruption.

Competition is widely used in the construction industry as a means of procurement. Design competition is primarily used when the conceptual quality of the design solution is paramount. This applies both to private emblematic buildings and to representative public undertakings, although this is used variably by public institutions in the European countries. Nevertheless, competitive tendering is the most commonly used means of selecting suppliers of construction services in Europe. It is actively promoted by many governments in pursuit of transparency, notably under the public procurement directives of the European Commission, mainly Directive 2004/18/EC on the coordination of procedures for the award of public works contracts, public supply contracts and public service contracts (European Commission, 2004). Competitive tenders may be open or selective on the basis of a pre-established tender list or standing tender list. The advantages of competition are that it:

- generates opportunities for fresh competitors
- brings transparency to the procurement process
- encourages production efficiency.

Against these advantages, there are some important disadvantages:

- only limited information on competitors is available for decision
- some less important aspects of the offer may be over-assessed to the detriment of possibly more important issues (particularly in design competition)
- procurement costs are high, both to the owners (rewards, search and selection) and to respondents to the competition (in order to prepare the offer, which may include creative work as in design competition)
- requirements for high uncertainty transactions are difficult to prepare
- underestimated winning tenders may lead to motivation problems during contract delivery.

In view of the above, it may be concluded that the procurement approach plays an essential role in building the project team. The most common procurement options in European countries are: traditional procurement, design-build arrangements, management contracting, construction management and public–private partnerships (PPPs). The following sections of this chapter discuss each of the above procurement options and describe the role of the main project team members.

Box 3.1 illustrates a case study highlighting the advantages of the use of an electronic platform in e-procurement.

Box 3.1   Vortal, a case study of an electronic platform

Computer-based procurement tools have been extensively developed in Europe over the last few years. Actually, e-procurement, e-tendering and e-submission platforms are now in common use in most European countries. Vortal is a Portuguese B2B operator that provides this type of solution.

VortalGOV is an electronic platform enabling the electronic integration of public buyers and suppliers making transactions securer, more confidential, faster, transparent, easier and more effective. The platform allows for performing the whole procurement and contracting processes free from paper-based forms. VortalGOV allows access to the global

---

(Continued)

network, enabling companies to cross borders and go global for a fractional cost. The transactional area of Vortal platforms is available in English, Spanish, Portuguese and Czech.

E-constroi is another platform for the construction sector, the largest acting in the Iberian Peninsula, and a success reference in Europe. E-constroi is designed for all purchase and supply companies in the construction sector. It supports a wide range of functions enabling users to:

- concentrate the procurement and bidding procedures on the same platform
- reduce costs by simplifying, automating and speeding up the contracting process
- follow up procedures on a real-time basis
- obtain evaluation feedback from the purchasing entities
- be aware of their market share among E-constroi users
- acknowledge supplier payment days.

3.3   Traditional procurement

Traditional procurement assumes the separation of design and construction (Ashworth, 2008, 2012; CRC, 2008; Winch, 2010). Figure 3.1 illustrates the relationships under this procurement approach. Typically, the owner first contracts design either to an architect for a building project or to a consulting engineer for an engineering project, and they become the owner's agent during the design phase and act as the design team leader. In some countries, a quantity surveyor may be designated, but in other countries quantity surveying takes place within each design area, with the coordination between areas ensured by the design team leader.

For small projects, the architect or the consultant engineer, as the case may be, can take the role of contract administrator on behalf of the owner. Alternatively, a member of the owner's staff can be appointed for this. In recent years, the figure of project manager has become common in many large traditional contracts, acting on behalf of the owner and as the owner's representative. The duties of the project manager vary according to the owner's needs and to the stage of the project life-cycle they are appointed for. Most professionals argue that nomination at the inception stage is beneficial for the project, but in most cases this only happens some time later, even for the construction phase only.

Coordinators may be appointed to embrace the responsibility and authority of part of the project team (e.g. the design team, pre-construction team or the construction team). Moreover, this is compulsory in some European countries for the design team and the construction team. The design coordinator is the design team leader and is an architect or an engineer, depending on the type of project. Construction phase coordinators are often called project directors, supervisors, inspectors, contract administrators or resident engineers (for civil

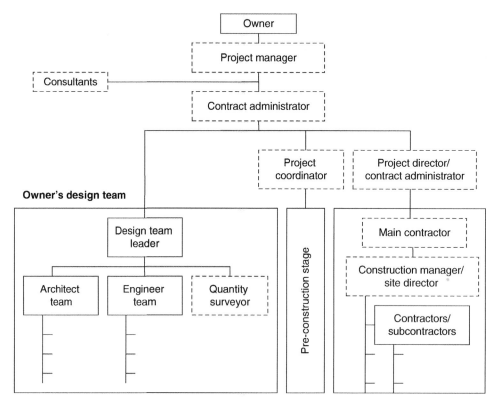

Figure 3.1 Traditional procurement relationships.

engineering contracts). The duties of the project director include inspecting and controlling the works, supervising payments to the contractor and clarifying design questions.

Under the traditional procurement approach, contractors are approached at the pre-construction stage; eventually, a contractor is selected to carry out the works, following the procedures mentioned in the previous section of this chapter. In some European countries, a main contractor must be nominated, but in other countries various contractors may be appointed to perform distinct parts of the works, especially in private projects. In the latter case, the contractor administrator or co-ordinator becomes essential for ensuring coordination.

The main contractor must nominate a representative acting on the contractor's behalf, who tends to be called construction site manager, site director, contractor's agent or even project manager, depending on the countries (the last title should not be confused with the owner's project manager discussed above). Construction works on site usually take place under the direction of the construction site manager who has the responsibility and authority assigned by the main contractor. The main contractor employs its workers on site and is normally allowed to subcontract part of the job to other companies (specialist subcontractors or labour-only subcontractors). The extent to which this may be done depends on the national law and on the contract conditions; these may impose limitations on the amount of work to be outsourced and may restrict eligible subcontractors to those approved by the owner at the tender stage, after the contractor's proposal (see Chapter 11 for details).

The owner's consultants may be of several types (legal consultants, financial consultants, etc.). Additionally, the project manager, if assigned, may have their own consultants, including quantity surveyors, planning supervisors, facility managers and so on.

3.4 Design-build arrangements

Design and build arrangements aim at overcoming some of the criticism of traditional procurement, namely the responsibility chain, the need for complete design prior to signing the contract and too late involvement of the contractor in the project life-cycle (Alhazmi and McCaffer, 2000). This has been claimed to lead to frequent disputes with costly implications both for owners and contractors, and to poor project performance, notably, overspent budgets and delays (Molenaar and Gransberg, 2001; Gransberg et al., 2006; Chan et al., 2011).

The design-build procurement method has many variants, depending on the owner's role in the process. Under the owner-led design-build approach, the owner opts to be considerably involved in the design. Accordingly, the owner first contracts the design up to a specified level of development. Currently, only the conceptual or at most the preliminary design will be contracted, but sometimes the owner wishes to carry out the design further than that. Contractors will be invited on the basis of the design documents developed so far. At the other extreme, there is the contractor-led design whereby the contractor is invited on the basis of the outline brief and is responsible for the development of design from that point (Figure 3.2). Accordingly, design-build arrangements acquire several names in current procurement practice, depending on the proportion of design undertaken by the owner prior to launching the tender (Cooke and Williams, 2009; Winch, 2010; Ashworth, 2012).

Design-build tender documents usually request the contractor to submit some design development from the design documents included in the tender inquiry. In mainstream construction practice, owners take design to the conceptual stage, in order to ensure their essential options, and request contractors to submit preliminary design documents at the tender stage, but other options are viable as well. The owner may possibly decide to conduct the whole process directly, but in most cases a project manager will be appointed. The project manager acts as the owner's representative and advisor in cost issues, and so in most cases keeps the quantity surveying role throughout the full process.

In order to develop the design, the contractor may use in-house capabilities or outsource design to an independent team. Theoretically, two design teams

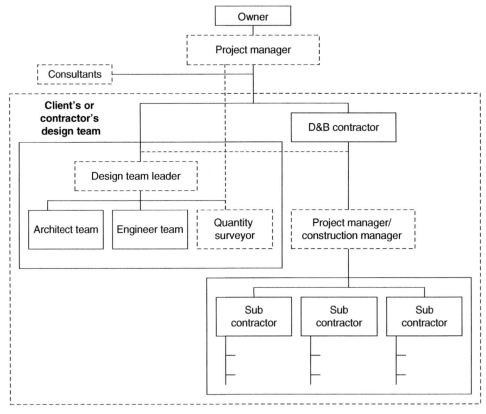

Figure 3.2 Design and build procurement relationships.

will be involved in this type of contract: the first one acting on behalf of the owner, before launching the tender, taking design as far the owner decides convenient; the second one, being employed by the contractor, developing the design for tender purposes and taking it further when the contract is awarded. In some cases, however, the owner may wish to give the responsibility of completing design to the first design team although now acting on behalf of the designated contractor. This may be interesting to the owner to give confidence that the project's goals will be achieved, but may raise ethical problems for the professionals involved.

3.5 Management contracting

Figure 3.3 shows the procurement relationships in a typical management contract. Under this approach, the owner appoints a person or a company to manage a set of work package subcontractors performing the construction works on site (Murdoch and Hughes, 2007; Brook, 2008; Cooke and Williams, 2009; Winch, 2010; Ashworth, 2012). Additionally, the management contractor

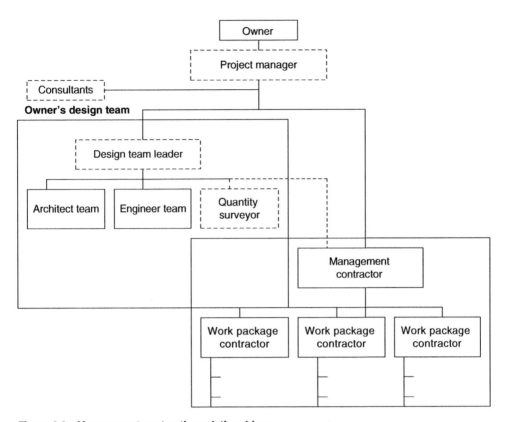

Figure 3.3   Management contracting relationships.

cooperates with the owner's design team during the design stages and in planning the construction works. This approach allows for overlapping design
and construction because each work package may be launched only after the
corresponding design is sufficiently developed. This may result in reduced
contract period and lower construction costs.

Accordingly, the owner first contracts the design and then contracts the
management contractor. This allows the owner to keep control of the design,
while benefiting from the experience of the management contractor (e.g. constructability and value engineering) during the development of the design.
Work package contractors are appointed on the owner's approval and contracted by the management contractor. Work package contracts may include
design of some project sections not assigned to the design team. For these
work packages, the procurement approach is identical to the one discussed in
the design-build approach. The management contractor works on the basis of
a fee covering professional advice to the design team, the preparation of the
project execution, the planning, programming and cost estimating of the
work packages, the procurement of work contractors, the management of
the construction works on site and so on.

Management contracting has several variants. For example, the owner may accept that the management contractor may enter in the competition for a specific work package. In case of being awarded a work package, the management contractor employs their own resources to fulfil it. Alternatively, the owner may require the management contractor to offer for some work packages if design is already sufficiently detailed for those packages (e.g. foundations and structures); in this case, the procurement approach is traditional for those packages. In other cases, the owner may wish to contract directly some work packages (e.g. those more specifically related to the owner's activity) but still demand the management contractor to handle the works. In the limit, the management contractor could just hold the responsibility of managing all work packages contracted by the owner, under a procurement approach known as construction management (see next section).

3.6   Construction management

The differences between the management contracting approach and the construction management approach may be seen by comparing Figures 3.3 and 3.4. Under the construction management approach, the owner directly contracts each work package, and the construction manager prepares the

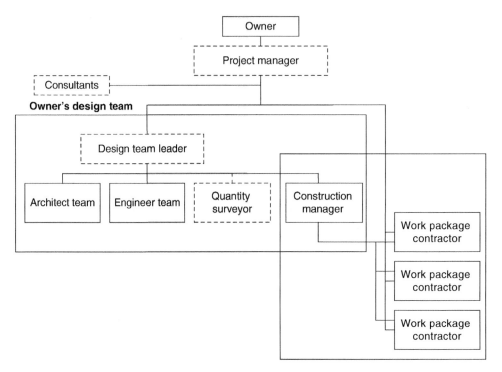

Figure 3.4   Construction management relationships.

corresponding project execution and coordinates works on site. The project manager normally carries the procurement of work package contractors (Murdoch and Hughes, 2007; Brook, 2008; Cooke and Williams, 2009; Winch, 2010; Ashworth, 2012).

Construction managers are usually appointed at the early stages of project development and integrate the design team early in the project because of a stipulated fee covering their duties. This function should not be confused with the construction site manager function whose functions have been mentioned above.

3.7   Relational contracting

Procurement models described so far reflect the conventionally adversarial relationships between construction project stakeholders. Although the credit for those models has been mostly acquired through the success of controlling the opposing interests of project participants, cooperative or relational approaches have long been practised among various construction actors. Examples of these are partnering and supply chain management practices, particularly when the adversity of the construction environment or the complexity of a specific project demand a joining of efforts between participants. The aim of relational contracting is expanding the benefits of the relational approach to all stakeholders in a construction endeavour.

Relational contracting is therefore a contracting approach based on the significance of cooperative relationships between project parties. Relational contracts are mostly used in complex long-term projects, taking place in contingent environments, usually requiring substantial relationships between parties for achieving project success. Additionally, relational contracts usually allow some flexibility in order to enable some adjustments if necessary. However, the scope of relational contracting may be broader than a single project, and the effects of it may extend beyond project boundaries (College, 2005).

Relational contracting embraces various approaches, such as partnering, joint venturing, alliancing, integrated project delivery and other collaborative working arrangements and risk sharing mechanisms (Masterman, 2002; Rowlinson and Cheung, 2004; Cheung et al., 2006).

Partnering is a process whereby team participants cooperate in improving their joint performance. The Construction Industry Institute (CII, 1991; Cooke and Williams, 2009) classifies partnering into three types: experimental partnering, packaged partnering (typically a client/contractor relationship) and committed partnering (embracing several stakeholders and supported by a firm dispute resolution mechanism). Partnering has also been categorised into project partnering and strategic partnering. The former is project specific and terminates with the project, while the latter mostly takes place at company level on a long-term basis and is applicable to various projects or construction activities. Actually, partnering has existed for a long time as an efficient approach to collaborative work between two or more construction actors (e.g. a design firm and a contractor, material suppliers and contractors, general contractors and subcontractors) or even between clients and construction companies (there are several examples over the past centuries). However, the latter form of partnering has gained considerable attention in the past few years, and some European countries have been formalising partnering procurement practice. Public-private partnerships (PPPs) are current contract arrangements, binding one or more public client and various construction providers, typically associated into a company specifically devoted to delivering a construction project facility (Akintoye et al., 2003). This type of partnering will be further discussed in the next section of this chapter.

Typical construction joint ventures are association agreements between construction suppliers to perform a specific contract. Accordingly, joint ventures are identical to project partnerships and are treated equally in law in most legal systems, but they may entail a broader scope and may also become similar to strategic partnerships. Joint ventures are often referred to as either populated or unpopulated, depending on their possessing staff of their own or using partners' staff only. Another approach for joint venture implementation is to set up an independent entity, usually a limited liability company.

Unlike partnering, where parties remain independent throughout the relationship, alliancing entails establishing a joint entity that shares risks and rewards according to an agreed model. Similarly to partnering, however, alliancing is also categorised into project alliancing and strategic alliancing, according to the definitions given above. Alliances involving two or more entities are common nowadays in construction but seldom include clients.

Integrated project delivery (IPD henceforth), is a collaborative alliance of professionals, business structures and practices into a delivery system that seeks to optimise project results while aligning member's interests through a team work approach (Ballard and Howell, 2003). The idea behind IPD is harnessing the abilities of team members to meet project aims and to increase its value for the client. This resembles the design-build approach described above, whereby designers and builders join in a team for project delivery. Differences lie in the stronger links between IPD team members and in the closer relationship with the client than usually happens under the design-build arrangements, and this results in less adversarial relationships between team members and with clients than in the former approach.

The implementation of IPD requires the adoption by the members of efficient information and communication systems. Although traditional information technologies may still be used, they may become quite time-consuming as the project complexity increases. Therefore online construction collaboration technology has acquired a relevant role under IPD, namely through the use of building information modelling (BIM), as explained in Chapter 4. Box 3.2 introduces BIM as a key tool for an IPD approach.

Box 3.2   IPD and BIM

A building information model (BIM) is a building product model containing all the information regarding the building's design and the processes taking place during the complete life-cycle. Accordingly, the model acts as an intelligent data source to store information for all disciplines involved. By merging all this information, BIM supports a better understanding of the project, drives collaboration between project participants and provides the basis for the implementation of integrated project delivery in construction. Regarding the software technology, a BIM consists of two major components: a three-dimensional digital representation of the building's geometry allows for its full visualisation and design; and a relational database stores all the data, properties, relations and performances. Because database and geometry objects are linked, all information stored can be referred to specific building components in the 3D visualisation. Examples of applications are design process visualisation, access to design calculations, information on material properties, cost estimating, scheduling, access to legislation applicable, constructability analysis and facility management support. The main advantages of BIM for major project stakeholders are as follows:

- owner: improved project understanding and support for the decision process
- design team: greater collaboration between team members; consistent project documents and standards to be used project-wide by all stakeholders; better detection of project inconsistencies and building conflicts
- contractors: improved tendering process; enhanced cost predictability; reduction of errors, omissions and less reworking; better construction planning and progress control
- authorities: improved checking of normative requirements; simplification and acceleration of the design approval process
- manufacturers: easier material ordering and less specification non-compliance
- facility managers: improved checking of building operation data against legislation and standards on energy, environmental impact and sustainability; easier programming of building maintenance, replacement and adaptability of building components.

In recent years, new forms of procurement have been developed in several European Union countries, because the previous forms have proved to be inefficient or defective in various aspects, especially when it comes to risk sharing. Framework contracting and prime contracting are only two of these new forms, basically introducing alternative ways of selecting the team.

The idea is to use an identical approach to the traditional standing tender list, while building more reliable and lasting project teams. Contracting for a specific project may, however, follow one of the forms described in the previous sections, thus contract relationships are identical.

3.8 Public concessions and public-private partnerships

A concession is a business operated by a company (the concessionaire) under a contract with, or a license from, the concession company. Nowadays, concessions are common arrangements for private commercial areas (department stores, sporting arenas, etc.), public spaces (e.g. a vending kiosk on a public square), service provision (e.g. catering services) and so forth.

A public service concession is an agreement between a private entity and a public organisation that grants the former the exclusive right to operate, maintain and carry out investment in a public utility for a given time period. Public concession contracts comprise a number of variants according to the rights and duties of parties involved. Leasing, for example, excludes private investment, which remains under the responsibility of the public entity; under management contracting the concessionaire operator only collects the revenue on behalf of the public concession entity and is paid an agreed fee. Public-private partnerships (PPPs) discussed in the following paragraph may also be considered similar to concessions.

PPPs are contract arrangements, whereby public sector investments are acquired through private sector funding. The idea behind this model is reducing public sector borrowing and fully involving the private sector in the life-cycle of large public projects, including financing, designing, constructing, operating and maintaining. Accordingly, there are several types of PPP schemes in practice, depending on the extent of private involvement in the project (Akintoye et al., 2003): the design-build-operate (DBOP), the design-build-finance-operate (DBFO), the build-own-operate-transfer (BOOT), the design-build-own-operate-transfer (DBOOT) and so on.

Typically, under this type of arrangement, the public sector sponsor first establishes a business case strategy and then launches a prequalification tender for selecting a short list of bidders. Bidders for PPP contracts are usually joint ventures or consortia of several companies (designers, construction companies, equipment suppliers, facility managers, etc.) organised into a special purpose company (SPC), backed by funding entities (banks or investment companies) and by insurance companies, plus a myriad of consultants (financial consultants, lawyers, technical experts, etc.) merged into a complex contract-funding-insurance relationship mesh. Figure 3.5 shows a simplified PPP relationship for a DBFO contract.

PPP arrangements are risky for its participants for a number of reasons. First, the consortium may fail, bringing losses for its participants and to the owner. This is the reason for insuring the operation and for the funding

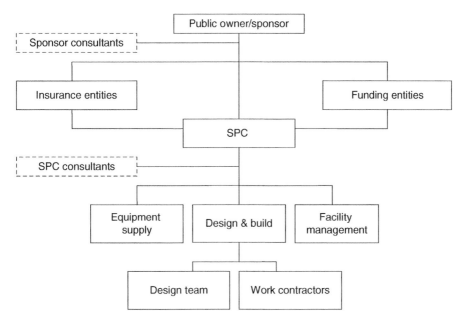

Figure 3.5   PPP relationships for a DBFO contract.

(Continued)

available will allow for the accommodation of roughly 150 people. According to the project aims, the accommodation should be designed mainly for students, researchers, visiting teachers, young professionals and others who may have a connection with local higher education institutions and may need accommodation for some time in the city.

In order to implement the project, Porto Vivo SRU decided to embark on a public-private partnership for a DBO contract. Following the Portuguese Code of Public Contracting, the first step was to launch an open tender to select the private partner. The consortium selected could benefit from a set of incentives conceded for private investors and further apply for support from the Portuguese Framework Programme (QREN).

According to the contract terms, the contractor holds the responsibility for designing, building and operating the premises for a period of 50 years, under the concession of Porto Vivo SRU. As compensation, the private consortium will pay 12.5% of the operational results before taxes (this figure is changeable according to the support obtained from QREN). The initial investment agreed is about €4.7 million and the maintenance costs were estimated at €315 000 every ten years.

The duties of Porto Vivo SRU are freeing the buildings and conceding the spaces to the private partner for implementing the project and operating the built facility. Porto Vivo SRU is also responsible for managing and accepting the project design, conducting the process of approval and construction authorisation, monitoring construction and following the first year of operation. Finally, Porto Vivo SRU is responsible for settling all problems emerging from the effects of the construction activities in the public space and in neighbouring properties (provisional use of public space, effects on access rights of third parties, traffic disruption and diversions, etc.).

Beyond designing, building and operating the student residence, the consortium also has a duty to obtain the corresponding permissions required by law. The consortium has to develop the project according to the schedule agreed in terms of the contract with the public partner, otherwise it may incur penalties and possibly contract cancellation. The construction period was stipulated as 18 months after obtaining all the necessary building permissions, and within 36 months after signing the contract with Porto Vivo SRU. Prior to opening the student residence, the contractor must have installed all the equipment and furniture needed and contracted and trained all the staff required for ensuring the operation and maintenance of the facility in adequate condition.

entities to compel the SPC to provide a minimum equity of the funds. Additionally, the owner may wish to ensure a step-in contract clause, enabling the project to continue in the case of the SPC failure. Second, the returns for the participants may be less than previously expected or they may even lose money on the project. Third, design contractors and work contractors have few warranties on their credits because they depend on a short-life company whose only assets are the project.

3.9   Organisation modelling

The organisation of the project team directly relates to the organisational structure of the entity undertaking it (Markham, 1998). Typically, organisation structures may be categorised into three basic models (Kerzner, 2009): functional, project-based and matrix-based. Projects undertaken by functional organisations are staffed by people from the same department within the organisation or else must be broken down into smaller portions if staff from various departments need to be involved. The advantages of the functional organisation approach for project management are sound reporting structures, clear authority (project managers tend to also be the functional managers), simple staffing, expertise of project members and previous relationships between them. But major disadvantages are lack of specialists within a department, lack of focus of project members (since they may not be allocated full time to the project) and delays in decision-making. The functional approach is adequate for small projects requiring staff from a few departments. Possible examples of this are maintenance works performed by contractors during the post-construction period.

For larger projects, the project-based organisation tends to be more appropriate. Under this approach, a project team is set up with people from various departments that are allocated full time to the project. Advantages include the generation of team attitude, comprehensible authority, clear focus of project members (since the project undertaking is their main concern) and possible resource redundancy (although the duplication of resources may work as a disadvantage, for cost reasons). Examples of this are current engineering projects conducted by large construction companies and complex design buildings developed by consultant firms.

The matrix approach seeks to merge the advantages of the two previous models while discarding their disadvantages. Accordingly, the matrix approach allows the staff from functional departments to focus on their specific working areas while supporting with specialists the projects running in the organisation. The project manager is appointed by the dominant department of each project or else by a specific project manager department within the organisation. The main advantage of this model is the efficient allocation of all resources, especially those skills that cannot be fully committed to one single project. The main disadvantage is that reporting relationships are complex. Examples of use of the matrix approach in construction are typical client in-house management teams.

However, the above models are rarely pure. For example, large contractors tend to work under mixed organisational structures; for example, the project management department is project based but the finance, human resources and purchasing departments are essentially functional. Another example is when the matrix model is used to a varying extent by organisations.

For a given project, the above comments are applicable to each organisation involved but for most construction projects, various organisations will participate. Therefore, the organisational structure of the project team is of paramount importance to the project performance. Figures 3.1–3.5 depict the current project organisational structures and show the links between participating organisations. The next section is devoted to discussing the project management team structure because of its central role in the project organisation's structure.

3.10   The project manager team

The project manager, acting on the owner's behalf, has a central role in the project organisation structure of most procurement models (Kerzner, 2009; Winch, 2010). Under the traditional procurement approach if the engagement terms of the project manager include contract administration (or project direction), then the project management team ought to include a supervisory structure for following design implementation on site and inspecting the works performed by the contractors; alternatively, the project manager may let the job to an external team, but under the project manager's direction. If the contract terms do not include contract administration, then the owner usually contracts a project director independently. Figure 3.6 shows an example of the project director team for a civil engineering project requiring laboratory testing.

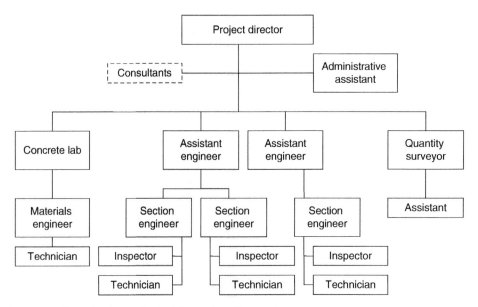

Figure 3.6   Organisational chart of the project director team.

Under the design-build model the contractor's agent appointed for managing the contract acts as contract administrator. Under management contracting or construction management models, it is the representative of the management contractor or the construction manager who acts as such.

The contractor's site organisation structure varies significantly from site to site because it is influenced by a number of factors, such as the type of construction project, the amount of work subcontracted, the existence of other projects being carried out in the vicinity, just to mention a few. In broad terms, however, the contractor's organisational structure is parallel to that depicted in Figure 3.6 headed by the construction site manager appointed by the main contractor. Chapter 5 explains the standard construction site management organisational structure.

References

Akintoye, A., Beck, M. and Hardcastle, C. (2003). *Public-private Partnerships: Managing Risks and Opportunities*. Blackwell, Oxford.

Alhazmi, T. and McCaffer, R. (2000). Project procurement system selection model. *Journal of Construction Engineering and. Management*, 126(3), 176–184.

Ashworth, A. (2008). *Pre-contract Studies. Development Economics, Tendering and Estimating* (3rd Edition). Blackwell, Oxford.

Ashworth, A. (2012). *Contractual Procedures in the Construction Industry* (6th Edition). Pearson, New York.

Ballard, G. and Howell, G.A. (2003). Lean project management. *Building Research & Information*, 31(2), 119–133.

Brook, M. (2008). *Estimating and Tendering for Construction Work* (4th Edition). Butterworth-Heinemann, Oxford.

Chan, D.W.M., Chan, A.P.C., Lam, P.T.I. and Wong, J.M.W. (2011). An empirical survey of the motives and benefits of adopting guaranteed maximum price and target cost contracts in construction. *International Journal of Project Management*, 29(5), 577–590.

Cheung, O.S., Kenneth T.W. and Chim S.P. (2006). How relational are construction contracts? *Journal of Professional Issues in Engineering Education and Practice*, 132(1), 48–56.

CII (1991). *In Search of Partnering Excellence*. Construction Industry Institute Special Publication 17–1, Bureau of Engineering Research, The University of Texas, Austin.

College, B. (2005). Relational contracting: creating value beyond the project. *Lean Construction Journal*, 2(1), 30–45.

Cooke, B. and Williams, P. (2009). *Construction Planning, Programming and Control* (3rd Edition). Wiley, Oxford.

CRC (2008). *Report on Building Procurement Methods*. Corporate Research Centre for Construction Innovation, Brisbane, Australia.

European Commission (2004). *Directive 2004/18/EC on the Coordination of Procedures for the Award of Public Works Contracts, Public Supply Contracts and Public Service Contracts*. European Commission, Brussels.

Gransberg, D.D., Koch, J.E. and Molenaar, K.R. (2006). *Preparing for Design-Build Projects*. ASCE Press, Reston (Virginia).

Kerzner, H. (2009). *Project Management: A Systems Approach to Planning, Scheduling, and Controlling* (10th Edition). Wiley, New York.

Markham, S.E. (1998). The scientific visualization of organizations: a rationale for a new approach to organizational modeling. *Decision Sciences*, 29(1), 1–23.

Masterman, J.W.E. (2002). *Introduction to Building Procurement System* (2nd Edition). Spon Press, London.

Molenaar, K.R. and Gransberg, D.D. (2001). Design-builder selection for small highway projects. *Journal of Management in Engineering*, 17(4), 214–223.

Murdoch, J. and Hughes, W. (2008). *Construction Contracts: Law and Management* (4th Edition). Spon Press, London.

Rowlinson, S. and Cheung, Y.K.F. (2004). A review of the concepts and definitions of the various forms of relational contracting. *International Symposium of the CIB W92 on Project Procurement for Infrastructure Construction*, 7–10 January, Chennai, India

Winch, G.M. (2010). *Managing Construction Projects* (2nd Edition). Wiley, Oxford.

Further reading

Ashworth, A. (2012). *Contractual Procedures in the Construction Industry* (6th Edition). Pearson, New York.

Murdoch, J. and Hughes, W. (2008). *Construction Contracts: Law and Management* (4th Edition). Spon Press, London.

4 Communications, Information and Documentation

4.1 Educational outcomes

Good communications between the parties involved, a steady flow of information, and adequate documentation of the activities developed during the construction project are key factors for analysing the tasks carried out and for avoiding later conflicts, considering that the goal of the company is the integration of processes at the project and business level. By the end of this chapter readers will be able to:

- identify the daily logs, reports, photographs and videos, construction diary and meetings as essential elements in providing adequate information and documentation
- outline the flows of information and documentation
- understand the importance of ICT in construction
- discover building information modelling and enterprise resource planning as integration tools in the construction phase.

4.2 Importance of communications, documentation and information

Until recently it was common for construction projects to take place with the minimum exchange of documentation possible among parties, in the belief that minimising records achieved a better working environment and improved the ability to meet objectives. This belief is flawed due to the fact that detailed documentation of events (both positive and negative) which take place during

Construction Management, First Edition. Eugenio Pellicer, Víctor Yepes, José C. Teixeira, Helder P. Moura and Joaquín Catalá.
© 2014 John Wiley & Sons, Ltd. Published 2014 by John Wiley & Sons, Ltd.

the construction project, offer multiple advantages (Fisk and Reynolds, 2009). It can:

- display the exact communication among parties involved, detailing record of the events and avoiding misunderstandings over the course of the project
- aid in controlling resource performance, measuring progress in such a way that it can be compared with estimated levels, analysing the causes behind deviations and taking measures to deal with the situation
- summarise the history of the project, allowing the preparation of conclusions and lessons learned for later actions
- justify decisions made on problems which may appear later, during the operation phase.

It is essential to prepare a document registry of the construction project, compiling all necessary data and ensuring access to all team members (Fisk and Reynolds, 2009). The document registry should cover aspects relating to events, conversations, correspondence, meetings and so on. It may contain: contracts, bids, permits, budgets, invoices, delivery notes, work schedules, data provided by the client, notes and drawings related to the design, technical specifications by suppliers and subcontractors, schedules, daily forecasts, reports, diaries, records, certificates, photographs, videos, etc. The transmission method may be via conventional post, fax, email, intranet, etc., including telephone and video recordings. A good document registry requires that the information be precise, objective, complete, legible, opportune and recoverable. The use of standard formats facilitates this work enormously (Civitello and Levy, 2005).

Remember that every contract (the main contract as well as those signed by subcontractors) requires that all requests, indications, deliveries of additional documentation and changes be issued in writing such that they can be recorded; the classic handshake or gentleman's agreement is not enough. The registration and documentation process should be carried out in such a way that they are accepted as reliable sources by the parties involved, particularly in case there is any type of later legal claim. In this case for example, it would not be sufficient for the project manager to keep notes of events without informing the contractor (Mincks and Johnston, 2010). This does not mean, of course, that the contractor must be in agreement with the record, but that the construction site manager should at least be advised. In addition, it must be adequately referenced, in order to ensure traceability, detailing if the document (e.g. a drawing) is being replaced or cancelled.

In addition, the storage of all important project material in a common and organised place facilitates teamwork and allows traceability of all tasks carried out in preparing it; this issue applies not only to paper documents but also to digital files (Fisk and Reynolds, 2009). Also, it avoids problems related to the loss of information and it also documents the history of the project. Figure 4.1

1. Contracting	2. Administration	3. Monitoring
1.1 Administrative specifications sheet 1.2 Technical specifications sheet 1.3 Company bid 1.4 Contract 1.5 Claims and disputes 1.6 Contract modifications 1.7 Invoicing 1.8 Payment 1.9 Guarantees 1.10 Others	2.1. Scheduling 2.2 Budgeting 2.3 Organisation 2.4 ISO 9001 quality 2.5 Changes 2.6 Subcontracting 2.7 Internal communications 2.8 External communications 2.9 Records 2.10 Permits 2.11 Design changes 2.12 Handover 2.13 Guarantee period 2.14 Final project documentation 2.15 Operation phase documents 2.16 Others	3.1. Daily logs 3.2 Photographs 3.3 Videos 3.4 Periodic reports 3.5 One-off reports 3.6 Geometric control 3.7 Quality control 3.8 Resource control 3.9 Productivity 3.10 Health and safety 3.11 Tests and inspections 3.12 Project closeout 3.13 Others

Figure 4.1   Example of file classification for a construction project.

provides an example of file classification, differentiating contracting, administration and monitoring; but these files could be combined in any other logical form. However, it is recommended that the filing and registration system be a standard system used by the company in every construction work. This way the need to create a completely different system for each project is avoided. Sometimes it may happen that the owner (through the project manager) implements their own document storage system; in this case both systems can be combined (the company's and the owner's), or alternatively the owner's may be used exclusively.

Archives are commonly used to store the information on paper, where all materials needed by each team member are organised, independently of any other general storage system which may exist. The same happens with the ICT files: these should be centralised in the company's intranet and correctly organised by place and time (Harris et al., 2006). Regarding the kind of document registries used during the construction phase, the following should be highlighted: daily logs, reports (periodic and one-off), construction diary, meetings records and photographs and videos. These are detailed in the following sections.

4.3   Communications on site

Communications play an essential role in conducting the project, both within each organisation team and between them. Good communications require not only speed and clarity but also definite channels through which information can flow. The approach to communications on a site is dependent on many factors: the nature of the project, the size of the contract, the organisation chart adopted by the project manager and the contractor, the command chain within them, and so on (Mincks and Johnston, 2010).

Additionally, the adequate process for communicating varies according to the nature of the subject and the formality of what information is being conveyed or what is being discussed. Requests, for example, tend to be quite formal, although informal communication may possibly be used in the first instance (e.g. approval of completed work). Rejection of work being carried should be done as soon as possible, but must be formalised quickly afterwards (Civitello and Levy, 2005). The agents involved in communication need to be selected according to their role in the process and to the level at which they want to place their messages. It may be inappropriate for the construction site manager to give an order directly to a labourer rather than through the foreman, following the command chain. Similarly, it is inappropriate for the representative of the owner to address a communication to the subcontractor without the main contractor being present.

Communications is a key factor for large working teams, which is usually the case in construction companies. It can be done electronically through email or through the company's intranet. It is used to provide written records of any circumstances or details which arise over the course of the construction project, or alternatively to give orders to the project team that are better specified in writing. In its most elementary format, it includes the addresses, sender, date, subject and a brief text section clearly and concisely written (Civitello and Levy, 2005).

Most of the communication on site actually takes place through discussions, telephone conversations and spoken messages, but they often need to be formalised in writing. This applies to requests, reports, proposals, approvals, rejections, technical notes and such like. Putting these in writing ensures that they are preserved for future reference, but they need to be clearly produced to have value (Fisk and Reynolds, 2009). There are many formats for communications between the contract parties on site; many contractors and owners have developed specific procedures for issuing adequate communications (see Figure 4.2 for a simple form for recording an informal meeting at the site or a telephone conversation). In recent years, ICT has allowed easier and faster means of transference, such as email, intra- and extranets, flash disks, DVDs or CDs (Chen and Kamara, 2008). The reason for this transference is preventing false interpretations of what was meant to be said or demanded. The style used in site correspondence is also important in the sense that it must be able to state what is needed without affecting the relationships between the parties concerned.

Irrespective of the need of unidirectional written communication, there is no substitute for agreed comprehensive statements on the matters addressed. In the current construction practice, this can only be achieved after open discussion between the parties who eventually agree on a written text adequately addressing the problem. Therefore meetings are the most adequate means of promoting discussion among project participants (Fisk and Reynolds, 2009).

Project:	
Day:	Time:
Record of:	**Ref.:**
Informal meeting ☐	Telephone conversation ☐

Parties involved:	
Owner ☐	Contractor ☐
Project manager ☐	Site manager ☐
Architect ☐	Subcontractor (specify) ☐
Other (specify) ☐	Other (specify) ☐

Subject:

Discussion:

Distribution to:

Signed by…

Figure 4.2 Informal meeting and telephone conversation form (developed from Civitello and Levy, 2005).

Informal meetings may be used for the less important subjects but may become unsuitable if topics discussed acquire more importance in the course of the discussion, because meeting participants may not have the authority to decide or may not be prepared to do so. In such cases, a formal meeting is then required. Unlike informal meetings that may take place at short notice or even just because people happen to meet at the construction site, formal meetings need to be arranged with reasonable notice to allow people time to prepare. Unlike informal meetings that may end up just in conversation, formal meetings are expected to be significant for the project.

Several types of prearranged meetings can take place on construction sites. Some of them are held at regular intervals and others are occasional for discussing particular subjects. For meetings to be useful, key persons relevant to the topic should be invited, a pre-established agenda should be followed and minutes should be produced (Fisk and Reynolds, 2009).

On most contracts, a regular progress meeting is held, usually at weekly intervals, for discussing current matters and to survey the project development (Mincks and Johnston, 2010). The project manager or someone from their team typically calls these meetings, and the contractor should also be represented by someone of equal status. Also, other people from the two teams may be invited, as well as representatives of subcontractors, designers and other external institutions, according to the topics on the agenda.

Beyond those, specially convened meetings also take place with some frequency on the construction site; the reason for this may be a specific issue that must be formally addressed (e.g. an incident, a claim or a design change), but it is decided not to include it on the agenda of the next regular meeting (e.g. because it may need an urgent decision, or it is too lengthy to accommodate in the normal meeting duration). Meetings with people and institutions external to the project also take place for a number of reasons. Examples are neighbours affected by the construction project, statutory undertakers, amenity organisations, local government departments with approval or supervision duties or the press. Summarising, a properly convened meeting, called at reasonable notice, with the right attendees and well prepared for the matters under discussion, working with a preset agenda, taking the time scheduled for it and producing correct concise minutes is a valuable instrument for adequately conducting the project (Fisk and Reynolds, 2009).

4.4   Daily logs

The daily logs detail all significant actions or important activities (formal or informal) that take place during the execution of the construction works (Figure 4.3). Specifically, the daily log should contain: log day and code, contract name, construction company, owner, construction site manager, weather conditions, resources employed (by subcontractors, indicating the level of manpower and equipment), supplies delivered (supplying company, quantity and storage location), in situ inspections carried out (tests, checks, geometry and risk prevention), consultations and meetings (including decisions made) and documentation exchanged, among others (Civitello and Levy, 2005). The log is usually filled in by the project team, signed by the construction site manager and submitted to the project manager.

These logs are issued daily even if no productive activity takes place. In this way, a project history exists which can be reviewed in the future for obtaining proper information (Mincks and Johnston, 2010). The logs can be issued on paper or in a digital version through email. The report should be limited to one single page, to minimise the time required to fill it in.

4.5   Reports

The reports can be regular or delivered on a one-off basis. Regular reports are usually issued on a monthly basis, although they may also be issued bimonthly, trimonthly or even weekly. A report is prepared which summarises all the work done during the relevant period (Mincks and Johnston, 2010). An outline of occasional reports issued for the period could also be included in this regular report. The focus of the report may vary if aimed at the owner (through the project manager) or the construction company. The information contained in the report could be different and would have to adapt to the circumstances. Even so, sometimes a single report can be prepared and sent to every party involved. The contents of the report are described in Box 4.1 (adapted from Civitello and Levy, 2005; Fisk and Reynolds, 2009; Mincks and Johnston, 2010).

DAILY WORKS LOG

DAY:
PROJECT:
CONTRACTOR:
SITE MANAGER:
OWNER:
PROJECT MANAGER:

REPORT N°:

WEATHER:
TEMPERATURE:
WIND:
HUMIDITY:

ON SITE RESOURCES:

COMPANY	MANPOWER	EQUIPMENT&MACH.	TIME ON SITE	ACTIVITIES UNDERWAY

SUPPLIES:

TYPE	COMPANY	TIME	QUANTITY	STORAGE

"IN SITU" INSPECTION:

TYPE	PART	CARRIED OUT BY	RESULT	OBSERVATIONS
TESTS				
CHECKS				
GEOMETRY				
SAFETY				

CONSULTATIONS / MEETINGS:

FROM	TO	SUBJECT	DECISIONS MADE

DOCUMENTATION:

FROM	TO	TYPE	SUBJECT

Signed by....

Distributed to....

Figure 4.3 Example of daily log.

Box 4.1 Contents of the report

- general description of the construction works: the activities carried out during the corresponding time period are described, commenting on areas worked, pace of performance, climatic influence, machinery available, land acquisitions, disrupted services and so on
- specific description of each of the important tasks, commenting on those that have begun and those been completed. An appendix with an extensive graphic report is generally included which reflects the progress on the work carried out and the problems encountered during construction (Figure 4.4 gives an example of project progress photographs)
- general descriptions of the other work carried out in collaboration with the project manager for the relevant period, including drawings, calculations and so on
- project timeline detailing the work carried out on a day-to-day basis. Normally the daily logs are included in an appendix and only the most important information is reflected in the report: incidents, meetings, milestones and so on
- written documents issued or received

Figure 4.4 Examples of project progress photographs.

(Continued)

- quality control: describes the tests and inspections carried out during the period, the work units affected and the results obtained. Where signs of non-compliance appear, the causes are analysed and corrective measures implemented. An appendix of test records and certificates is usually attached
- environmental impact: comments on compliance with the environmental plan, incidents that have occurred and their proposed solution. The minutes for environmental meetings held, and orders issued, are also included
- health and safety: details the main measures taken, incidents that have occurred and the proposed solutions. The records corresponding to minutes and orders are also included in the report
- compliance with the schedule: the construction schedule is analysed, commenting on gaps in the initial plan and their causes: weather, facilities, resources, services, changes, accidents and so on. The estimated schedule for the next period is specified and its feasibility is analysed
- economic valuation: first, the certified work is compared with the estimated progress for the corresponding period (Figure 4.5) then the real cost is compared to the predicted cost (Figure 4.6). On many occasions gaps are analysed in sections, bearing in mind the contractual budget, the consumed budget, the estimated final budget and the deviation of the contractual budget from the estimates. The deviation could be due to measurement differences, unplanned or unexecuted units.

Economic valuation summary of chapter A	
Description	Sum (€)
Contract budget	100 000
Consumed budget	77 000
Final estimated valuation	109 000
Deviation from budget	**+9 000**
Measurement difference	+4 000
Unplanned units	+7 000
Units not executed	−2 000

Figure 4.5   Example of economic valuation summary.

Cost Summary of Chapter A	
Description	Sum (€)
Contract Budget	100,000
Estimated Cost	96,000
Real Cost	75,000
Final Estimated Cost	99,000

Figure 4.6   Example of cost summary.

The reports can also be simple, displaying graphically the main information related to price and time frame:

- consumed budget vs. approved budget (original contract or subsequently changed): absolute, relative and total value
- consumed schedule vs. approved schedule (original contract or subsequently changed), including the partial milestones pending completion.

In addition, all information relating to the project team's actions is stated: incidents, unforeseen circumstances, problems, proposals for changes and so on.

4.6   Construction diary

The construction diary (or orders book) registers orders, communications and instructions from the project manager to the construction site manager, in addition to specifying all problems and incidents that could arise during the construction works (Mincks and Johnston, 2010). The project manager may also issue written documents to the contractor with orders or directions which are attached to the diary. The project manager's signature or that of any other authorised member of the project management team is needed, in addition to the confirmation of receipt from the construction company which confirms that the information has been accepted and a written copy received. The book is property of the owner. For public tendering, the order book is dealt with by the administrative body responsible for the construction project and becomes its property once the project is complete, preceding the handover of the infrastructure. This may be fundamental when there is any type of legal claim, ensuring that the parties involved do remember the incidents which took place at the time.

The book should be appropriately bound in such a way that pages cannot be removed or substituted. They should be numbered in consecutive order, it being prohibited to skip pages or delete text. Notes are made by hand to avoid tampering and to prove authenticity. Every working day should have a corresponding entry, even if no work is done; this good habit does not generally apply where entries are only made in the book when problems arise, leading people to believe that the less written documentation produced during a project the better for all (Fisk and Reynolds, 2009).

The diary should contain three specific key topics:

- problems detected, either a defect or omission in the project, bad execution of the construction works or other exceptional circumstances
- proposed solutions and recommended changes – the decisions made are written down
- orders transmitted which may affect an isolated case or be general, giving rise to changes in the technical specifications or the drawings.

If additional tasks must be carried out due to the appearance of unforeseen circumstances, then the resources available on the site, and the activities they are involved, should be detailed (Mincks and Johnston, 2010). Another important issue arises when the contractor is carrying out tasks or completing work that does not correspond to the design specifications. Any type of error made by any of the parties may also be reflected in this diary.

4.7   Meetings

Communication during the construction phase is generally conducted through meetings: internal (construction team) or external (between the parties involved, including subcontractors). The usual purpose of these meetings is the coordination of the construction works, providing a forum in which common topics of interest can be discussed between all the agent involved (Mincks and Johnston, 2010).

It is recommended that an agenda be prepared and sent to the attendees, both for internal meetings of the project team, and external meetings with the owner (or its representatives), subcontractors or other third parties, including the date and location of the meeting and the agenda (Figure 4.7). The meeting minutes are of vital importance (Figure 4.8), providing information with regard to the meeting date and location, the attendees, completion of pending actions, topics dealt with and decisions and agreements reached, specifying the actions to be taken, the responsible parties and the expected completion date. It is recommended that the date for the next meeting and its agenda are set prior to closing the session. The records of the minutes are distributed to all attendees following the meeting; and a time period is proposed in which the attendees can express their concerns regarding the contents of the minutes (Civitello and Levy, 2005). Corrections to the minutes should also be suitably registered and distributed to all attendees.

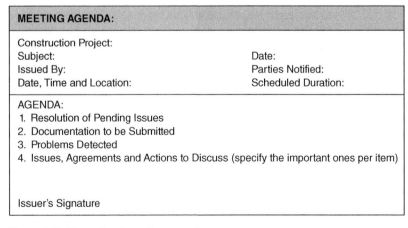

MEETING AGENDA:	
Construction Project:	
Subject:	Date:
Issued By:	Parties Notified:
Date, Time and Location:	Scheduled Duration:

AGENDA:
1. Resolution of Pending Issues
2. Documentation to be Submitted
3. Problems Detected
4. Issues, Agreements and Actions to Discuss (specify the important ones per item)

Issuer's Signature

Figure 4.7   Example of meeting agenda.

Meeting minutes	
Construction project: Subject: Minutes prepared by: Date, time and location:	Attendees: Duration:
Minutes: 1. Issues pending from previous meetings 2. Documentation submitted 3. Problems detected 4. Other issues 5. Agreements made 6. Proposed actions and the responsible parties (including due date) Attendees' signatures	

Figure 4.8   Example of meeting minutes.

The preliminary meeting with the project manager (owner representative) is one of the most important tasks before beginning the construction project. At this meeting the basic relationships between parties are established, and doubts concerning the design or other circumstances are discussed. All this information is compiled in a record of the minutes that the client could return with its approval; this does not always happen owing to reticence on the part of some clients (above all public administrative agencies) to provide written instructions (Mincks and Johnston, 2010).

4.8   Photographs and videos on site

On-site photographs are taken in digital format, which means that hundreds of photos can be taken daily. As indicated, the most relevant ones can be sent to the parties involved as an attachment to the daily log using email or intranet (Figure 4.4). Very few photos are usually included in the report on paper and they are more often than not the most representative. Nevertheless, due to the large amount of pictures generated, these should be adequately filed in such a way that the information can be easily retrieved using electronic systems. The best approach is to create a document database which allows searches to be carried out, not just by date, but by area, resources, employees, subcontractors or keywords. The photos can be linked to other key project information, such as daily logs, reports, meeting minutes or the project diary (Civitello and Levy, 2005).

It is very important to identify each photograph properly. As stated previously, the documentation generated by each project (including the photos) has to be stored in a specific file for that project. A sub-file can be named according to the day that the picture was taken, i.e. 'yyyy-mm-dd'; this way, the sub-files are ordered automatically by date. Additional information such as photo location (area and task) and orientation (direction of view) should also be kept. Taking photographs can serve to document the existing conditions prior to the construction works, including not only the site itself but also the adjacent property. This is very important when buildings or other structures can be affected in some way by the construction activities, such as earthmoving, pile driving or pipe jacking. Furthermore, it is also important to have graphical proof of elements which will later be hidden, as well as any changes or unexpected events (Civitello and Levy, 2005). Similarly video can also be used. This can also be used as evidence later in case of litigation or arbitration.

Another aspect recently gaining importance is the use of webcams (also called surveillance cameras) on site. These cameras transmit fixed images or low resolution video periodically from remote locations in real time. They allow authorised members (through a password) 24 hour access to the project through the company's corresponding website or intranet. The possibility of observing, monitoring and managing the project from a distance of thousands of kilometres has become a reality, allowing real-time viewing of the construction work being carried out. The cameras can zoom to allow details on site to be observed. In some cases, such as excavation, laser scan images can be used to obtain quantitative data also (Hashash and Finno, 2008).

4.9   Information and documentation flow in construction

Having formed the project team and determined the responsibilities of each of the members, a table or matrix of responsibilities can be prepared which clearly assigns functions to each of the team members. Then, an important step is to develop a document procedure which allows the filing of information (paperwork or digital) from different documents, including emails. In this way, the document can be easily generated or updated and could include a very powerful information recovery capacity, facilitating searches of data and analysis of information. This procedure also has to consider the automatic distribution of logs and reports to every party involved (in pdf or any other format); they can be accompanied by photographs or digital videos. Two key issues arise: the transmission of information and the distribution of documentation (Harris et al., 2006).

Figure 4.9 provides a representative example of the channels of communication and the frequency of transmission among members of the project team, the construction company, the subcontractors and the project manager (as a representative of the client). Every construction site manager should communicate daily with the field engineers, and weekly with their company superior and also the project manager, in their role as a representative of the client. Thier contact with the board of directors of the company or with the subcontractors is usually on a monthly basis.

DESTINATION						
ORIGIN	Construction site manager	Field engineer	Department manager	Board of directors	Subcontractor	Owner's representative
Construction site manager		Daily	Weekly	Monthly	Monthly	Weekly
Field engineer	Daily		Monthly	Quarterly	Weekly	As needed
Department manager	Weekly	Monthly		Weekly	As needed	Monthly
Board of directors	Monthly	Quarterly	Weekly		None	Quarterly

Figure 4.9   Communication matrix (developed from Pellicer and Yepes, 2007).

		Construction site manager	Field engineer	Department manager	Board of directors	Subcontractor	Project manager
D	Time scheduling	Source	Addressee	Addressee			Addressee
O	Cost estimation	Source	Addressee	Addressee	Addressee		
C	Subcontract	Source	Addressee	Addressee		Addressee	
U	Quality control	Addressee	Source			Addressee	
M	Daily logs	Source	Addressee				Addressee
E	Monthly report	Source	Addressee				Addressee
N	Technical document	Addressee	Source				Addressee
T	Final report	Source	Addressee	Addressee	Addressee		Addressee

Figure 4.10   Document distribution matrix (developed from Pellicer and Yepes, 2007).

Figure 4.10 is an example of a distribution matrix which reflects the origin and destination of documentation generated over the course of the project between the parties represented in the previous example. Some of the most common documents during a construction project were considered: time scheduling, cost estimations (internal documents), subcontracts, quality control, daily logs, monthly reports, technical documents and the final report.

4.10   Information and communications technologies (ICT)

Information is data that has undergone a transformation (or elaboration) process. Information becomes knowledge once it is interpreted, learned and reasoned in its context, facilitating decision-making and action-taking (Popper, 1972). An information system deals with three basic stages: identifying and capturing information, classifying information and disseminating information (Tushman and Nadler, 1978). This process generates problems such as: search, physical space for storing, reproduction, obsolescence and reliability. The effectiveness and efficiency of this process depend on the information gathered, and specifically on three main characteristics of that information: quality, quantity and opportunity (Björk, 1999).

The final decades of the 20th century witnessed an escalation of the development of ICT, as well as the implementation of these technologies in the construction industry (Fischer and Kunz, 2004). Every step of the facility lifecycle involves multiple interrelated activities, as well as many organisations and stakeholders. In order to optimise the life-cycle in general and the construction management process in particular, an integrated perspective is essential. ICT is a useful framework to meet this goal. There are technologies that allow the team and organisations to interact, using as much information as possible (Yu et al., 1999). Some authors (Vitkauskaite and Gatautis, 2008) still consider that the use of ICT is very limited in the construction industry, lagging behind other sectors, because of its specific characteristics, a predominance of SMEs and poorly oriented innovation. Other authors, however, believe that ICT is raising productivity through the quality and speed of work, financial controls, communications and access to common data (Onyegiri et al., 2011). Chen and Kamara (2008) have pointed out that construction sites, generally located far from the main offices and sometimes away from urban areas, used more traditional methods and did not have sophisticated ICT support, compared with the main offices, in which case there is a need to integrate information and communications between the work sites and the main offices, taking advantage of the ICT tools and improving productivity.

The first implementations of ICT in construction focused on very specific tools to help performing traditional tasks, mainly related to the design phase: structural analysis, hydrology and hydraulics, computer aided design (CAD), estimating, scheduling and so on, and even the typical office tools (word processing and spreadsheets); these tools are currently very mature, and several software products can be chosen from the market (Froese, 2010). In the mid 1990s, the scenario changed radically with the introduction of a global system of interconnected computer networks used for communication between computers, sharing and exchanging information and data (Onyegiri et al., 2011): internet, intranet and extranet computer-supported communications, such as email, document management systems, enterprise resource planning systems (ERP) and so on. This is a less mature field: there are still project and business processes that are still adapting to these tools. In recent years, individual applications on system integration at the project level have been developed, the most relevant of them being building information modelling (BIM) (Froese, 2010).

Figure 4.11 shows the different technologies and tools currently used in construction. From the software side, four approaches are considered: general, project and business. The general approach involves the traditional office applications that are common to every user in an organisation at every level: word processing and spreadsheets. In this same field there are the typical tools for written communications using email (e.g. Google Gmail™ or Microsoft Outlook™) or file storage and sharing (e.g. Dropbox™ or Google Docs™), and voice and image communications (e.g. Skype™ or GoToMeeting™).

The project approach displays three different lines depending on the target: specific discipline, design or management. Software for specific disciplines has developed since the 1960s: PLAXIS3D™ for geotechnical analysis of deformation and stability of soil structures, groundwater and heat flow; ETABS™ and SAP2000™ for design and calculation of building and other types

SOFTWARE

General	Project focused			Business focused
	Discipline oriented	**Design oriented**	**Managerial oriented**	
• Word processor • Spreadsheet • Email • Communications (voice and image)	• Soils • Structures • Energy • HVAC • Lighting • Pipelines • Others	• CAM • 2D & 3D CAD • Virtual reality	• Planning & scheduling • Cost estimating	• Human Resources • Accounting • Procurement • Contracts • Documentation • Communications • Others

HARDWARE

Computers	Networks	Peripheral devices	Communication devices
• Desktops • Laptops	• Internet (world wide web) • Local area network (LAN) • Wide area network (WAN) • Virtual private network (VPN) • Intranet • Extranet	• Printers • Scanners • Others	• Mobile devices • Fixed devices

DATABASES & DATABASE MANAGEMENT SYSTEMS

⇩ ⇩ ⇩ ⇩

INTEGRATION:

From the point of view of the project → building information modelling (BIM)

From the point of view of the organisation → enterprise resource planning (ERP)

Figure 4.11 Main applications of information and communication technologies in construction.

of structures, respectively; ANSYS™ for product simulation using finite element analysis; HEC™ software package developed by the US Army Corps of Engineers for hydrological and hydraulic analysis; TRACE™700 for designing HVAC system solutions; and so on.

Design-oriented software generally known as computer aided design (CAD), and its subsequent process, computer aided manufacturing (CAM), were first developed in the 1960s and 1970s by the automotive companies (Bozdoc, 2012). The emergence of personal desktop computers in the 1980s, as well as the foundation of Autodesk (AutoCad™) and Bentley Systems (MicroStation™), allowed the use of this software by small and medium companies, especially in the construction industry, in order to create a technical drawing with the help of computers. In succeeding years the boom of this tool reduced the numbers of draftsmen drastically in engineering and architectural firms, as well as in contracting companies. This software has evolved from 2D vector-based drafting systems to 3D solid models. Architectural and engineering rendering is the process of generating an image from a model using computer software. It is generally produced for presentation, marketing or design analysis purposes. CAD is currently overtaken and gradually replaced by BIM-oriented software. BIM will be explained in section 4.11.

Scheduling software tools are being developed since the introduction of the network scheduling (CPM and PERT) in the 1960s (Fisk and Reynolds, 2009). At that time, the characteristics of the computers did not allow for a massive use of these methods. However, the introduction of the personal computer in the 1980s favoured the purchasing of software by companies working in the construction industry. The most popular software tools are Primavera Project Planner™, launched in 1983 by Primavera Systems Inc. but currently owned by Oracle Co., and Microsoft Project™ launched in 1987. Both of these enable the use of bar and network charts; Primavera is better for large projects with many activities, whereas MS Project is more user-friendly.

There are many software tools that deal with estimating costs for bidding, planning and control in construction, but they are generally customised for the particular characteristics of each national market. Two general examples are: Bid4Build™ (buildings) or BID2WIN™ (heavy civil construction). These applications are commonly sold by the vendor with a supplied database including activities, definitions and unit prices; additional activities and prices can be introduced by the contractor. It is also possible to break down activities into several layers or phases of sub-activities as well as measurements (quantities).

Regarding the business-oriented approach, the company has to take into consideration several processes that need planning and control (Pellicer et al., 2009): budgets (including analytical and financial accounting), clients, suppliers (including subcontractors), communications (including document management), contracts (construction projects), resources (materials, equipment and machinery), organisational hierarchy, personnel and procurement (Figure 4.12). Software tools are available for many of these processes. The ones related to the accounting and financial systems of the company are

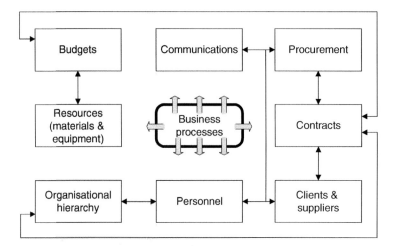

Figure 4.12   Main business processes in the construction company.

classic software packages (e.g. QuickBooks™, Sage Peachtree™ or GnuCash™). Other applications such as OpenDocMan™ and LogicalDoc™ focus on electronic document management, information and knowledge generated within an organisation.

Furthermore, the internet environment has allowed the introduction of the 'e-business' concept in the economy (Meena, 2009) – a virtual space in which economic exchanges occur through the use of ICT: B2B (business to business), B2C (business to customer), B2E (business to employee), G2C (government to citizen), G2B (government to business) and so on. Because of this e-business environment, the procurement process in construction companies has been made much simpler. Internet allows the identification of call for bids, the submission of bidding forms and proposals, information on public bidding (including bid opening and contract awarding), control of contract administrative procedures using the client's website, and the management of supplies and subcontracts through specialised websites too. Currently the business processes at the companies, also in the construction sector, are being integrated using enterprise resource planning systems (ERP); this is explained in section 4.12.

Hardware is the other important part of the ICT. It includes computers (desktops and laptops), networks, peripheral devices (printers, scanners, etc.) and communication devices (fixed and mobile). Networks combine hardware devices in different ways (Onyegiri et al., 2011). Apart from the global internet, local area networks (LAN) are high-speed communications systems that link together computers and other data-processing devices within a small geographic area, generally a company. Connecting several LANs creates a wide area network (WAN). Using one of those options allows a contractor to share information among the main offices and the construction sites, for instance. Accordingly, an intranet is a restricted access network, with features similar to that of the internet, but which is only available to an organisation internally; an extranet is an extension of a company's internal network that allows access to outside users through additional security and authentication (Onyegiri et al., 2011). Finally, a virtual private network (VPN) is an environment in which access is controlled to permit connections only to a defined group within the global internet.

A key basic element of the ICT field is the relational database, defined by Date (2003) as a collection of persistent data that is used by the application systems of the company. Persistent data means data that is not transient in nature and it can only be removed from the database using some explicit request. A database management system (DBMS) is the software that handles all access to the database (Date, 2003). A DBMS provides the company with centralised data control, and has several advantages: sharing data, reducing redundancy, avoiding inconsistency, supporting transactions, maintaining integrity and enforcing security (Pellicer et al., 2009). The key rule in a DBMS is that each piece of data can only be input once. Two kinds of data are put into the DBMS: primary data for configuration, which is introduced by the system administrator, and operative (regular) data that is introduced by the users.

As seen in Figure 4.11, the final goal of the ICT is the integration of software and hardware using databases and DBMSs. This is generally accomplished at two levels: project level through BIM; and business level through ERP. The former is explained in section 4.11, and the later in section 4.12.

4.11   Building information modelling (BIM)

Building information modelling (BIM) can be defined as the process of generating and managing construction information with an interoperable method that permits users to integrate and reuse that information and knowledge through the facility's life-cycle (Lee et al., 2006). As stated in section 4.10 (especially in Figure 4.11), not only design-oriented software, but also project- and business-oriented were developed separately in general until the beginning of this century. Integrated models were possible because of the development of object-oriented programming and parametric modelling (Ashcraft, 2007). Parametric modelling enables the translation of explicit geometric expressions, among others, which can automate the generation of the building information model (Lee et al., 2006).

Initially, the integration was done using 3D CAD plus time, defining the 4D concept using the parametric link (Fischer and Kunz, 2004); the model relates the different phases of the facility life-cycle, enabling the stakeholders to view the project over time and review the planned, actual and future status of the whole project or of any part of it. Some authors (Fischer and Kunz, 2004; Ashcraft, 2007) renamed the model as 5D when it also included cost information. BIM started to use 3D at its full power, expanding its usage into an nD

> ### Box 4.2   What are the advantages of BIM?
>
> - there is a single data entry, but multiple users, avoiding inconsistencies and errors
> - the database is unique, and all the parties share the same data
> - once the model is created, the design is more efficient, especially when changes start to be made
> - detailed (shop and fabrication) drawings can be extracted from the model by suppliers and subcontractors, reducing fabrication costs and errors
> - the 3D model can automatically detect physical conflicts between the elements, facilitating conflict identification and resolution
> - the model contains information related to cost, quantities and time, enabling bidding, estimation, control and productivity calculations
> - alternative options can be easily visualised
> - energy analysis can be performed, evaluating lighting design and generating documents for certification
> - the model allows simulation of the construction process, increasing the ability to evaluate and optimise it
> - once the construction is over, the model can be used to operate the facility, managing maintenance and remodelling, and evaluating emergency response and evacuation.

environment (Jung and Joo, 2011). However, this has been a slow but steady process in order to achieve multidimensional integration. 3D models attempt full-range integration adding an extra dimension that can be (Jung and Joo, 2011): engineering analysis in structures, energy and disaster prevention during the design phase, or planning, scheduling, control and safety during the construction phase.

BIM represents the design as objects which convey their geometry, relations and attributes (Eastman et al., 2011); these objects can be conceptual or construction detailed and parametrically defined. These parameters are contained in a database defining the attributes and the relationships between them; thus, a relational database is a key element of BIM (Ashcraft, 2007). When these objects are linked together in a rational way, they form a building model. Links between the objects can maintain or adjust elements of the model in response to design changes. The advantages of BIM are highlighted in Box 4.2 (adapted from Ashcraft, 2007).

BIM can serve, and is sometimes used, as the main tool for a collaborative framework, changing the relationships between the different parties. The use of BIM has benefits for owners and contractors (Ashcraft, 2007): better design optimisation and coordination and fewer errors, disputes and claims. The owner could exploit the model for the operation of the facility, using an as-built model. The contractor also benefits by the use of a more detailed model from which shop drawings and fabrication information can be obtained more easily, shortening the time to delivery. However, the benefits are not so clear for the designer, because they are the ones who have to make the investment, train the technicians and master its use. Furthermore, later the designer loses control of the design, even arousing conflicts of confidentiality and putting in danger the intellectual property (depending on the country); this might increase the designer's liability in case of disputes and claims. The rewards are asymmetrical (Ashcraft, 2007), hindering the adoption of BIM by designers. Some of these issues could be overcome using collaborative contractual systems, such as the integrated delivery method or other relational contracts explained in Chapter 3. Furthermore, these procurement approaches will increase cooperation between the parties involved, as well as integration of the full life-cycle of the facility.

4.12   Electronic business and project administration

Enterprise resources planning systems (ERP) are software packages (commercial or customised) that seek integration of the different processes that are carried out by the company (some of them are displayed in Figure 4.12). ERPs aim to improve business efficiency by providing accurate information on time to take optimum decisions while diminishing paperwork (Shi and Halpin, 2003). Effective management for a typical project-based construction company requires the availability of a flexible and easy-to-implement computerised information system specifically designed for this kind of company. The ERP has to deal with project-based and business-focused processes and has to be adaptable, working in real time and enabling the use of customised applications of common commercial software.

As explained in Chapter 1, the organisational hierarchy of the construction company defines the structure of the firm: board of directors, departments, sections, projects and so on. Each employee, through their work post, has a place in the hierarchy of the firm. Similarly, construction projects are appointed to the corresponding department, and a team project is chosen. Several different interfaces can be provided for each personnel category within the firm. The users are categorised according to their functions, differentiating between administrative or productive; a proposal of user categories by hierarchy is shown in Figure 4.13. Categories common to both types of staff are: system administrator, chief executive officer, head of department and head of unit. There can be specific categories for productive personnel, such as: construction site manager, field engineer, foreman and basic user. The access level to the different function of the ERP and the security acquired (authentication) depend on the location of the work post within the organisational hierarchy of the company.

Administrative staff	Productive personnel
	System administrator
	Chief executive officer
	Head of department
	Head of unit
	Construction site manager
Administrative personnel	Field engineer
	Foreman
Clerical personnel	Basic user

Figure 4.13   User categories by hierarchy (developed from Pellicer et al., 2009).

The ERP has to take into consideration the basic scheme of data flow in the organisation. It is illustrated in Figure 4.14, considering the following logical steps: data input, data validation, data storage and exploitation, output of information, and action. Every employee inputs time spent on activities regarding the corresponding project; expenses, such as subsistence, transportation, supplies and so on, can also be input. Employees working in network-equipped offices input data every day through the intranet, whereas employees working on site can introduce data using the intranet (through network-connected offices or via mobile devices) or by means of reports delivered weekly to the administrative staff.

On a daily (or at least weekly) basis, administrative personnel input supplies, rented equipment, outsourcing services and subcontracting to each contract through the computer system; Figure 4.15 shows a simplified interface for this case. Invoices are also recorded to the corresponding contracts. Also, administrative employees can input external data to be available for benchmarking purposes. Internal controllers (generally the upper person in command) may check the data input, whether they are administrative or productive. This supervision procedure can be softened or hardened according to work posts, functions, tasks and contracts, always depending on the corporate policy (Pellicer et al., 2009).

With all necessary data in the relational tables of the DBMS, useful information can be generated. Reports are specifically designed to obtain certain information related not only to cost and time (per person, task, project, department or company), but to bids and contracts (regarding their status, resources, budget, etc.) as well. A typical monthly report states company's own personnel, machinery and equipment time (input hours, theoretical hours and overtime), personnel subsistence costs (mileage, travel displacements, restaurant meals, others related to travel such as parking), costs of outsourced (subcontracted) personnel, machinery and equipment (as well as work planned vs. work performed) and, finally, costs of supplies (including quantities).

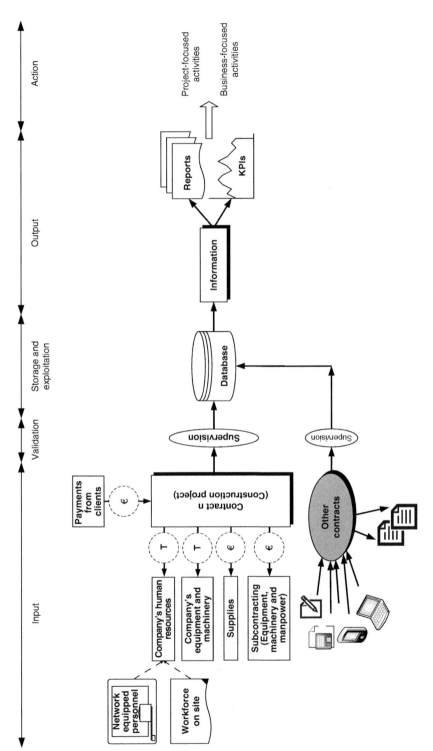

Figure 4.14 Information system for construction projects (developed from Pellicer et al., 2009).

Queries from the DBMS enable staff to analyse the variables that affect the performance of the firm, such as the yearly evolution of productivity and profitability, or financial ratios (Pellicer et al., 2009). The information allows comparison between planned and current status, or to estimate future scenarios. This is a crucial factor in order to avoid potential problems and to facilitate (internal and external) benchmarking. External ratios can also be input into the ERP and used for contrast.

Nonetheless, construction sites, as stated previously, are placed whenever the project requires and, therefore utilise conventional methods with less ICT support, for instance sheets of paper and field notes (Kimoto et al., 2005). There is a delay between writing the report and inputting the data into the company's system; this can sometimes produce misinformation and confusion. Mobile devices, such as personal digital assistants (PDA), small laptops, iPads, tablets or even smart-phones, can decrease this problem (Kimoto et al., 2005). This way, construction managers, superintendents and foremen can input digital data directly to the information system using mobile devices. Box 4.3 shows a case study of the implementation of an ERP system.

User's name	
Contract reference	
Task performed	
Time worked	
Other costs	
Comments (if any)	

Figure 4.15   Simplified ERP's interface for daily report.

Box 4.3   A case study of adaptation to ICT

At the turn of the millennium, a specialised SME contractor with 80 employees and an annual turnover of €20 million was still controlling its business in the traditional 20th century way. First, each employee filled out one form (on paper) per day indicating the work performed. Every day each form was collected by a clerk; this employee handed it over to the responsible person, who validated it. Later, the clerk took the forms back from the responsible person and returned them to the administration department. Someone in this department (an assistant) inputted the data in a spreadsheet, assigning times and costs per person, activity and project. This same person, or another, also inputted the invoices from suppliers and subcontractors.

The owner of the company (and also CEO) noticed that a lot of time was wasted in this process. He asked advice from his regular financial

(Continued)

advisor, an independent collaborator of the firm who then proposed that the owner implemented a customised ERP in phases. As well as the advisor's time, an additional part-time computer technician was also needed during the implementation period. The ERP was developed according to the needs of the company. The first phase developed the construction project module that was going to control the time and cost of every project at the company, integrating its results into the accounting and financial systems. Six months later, after some initial resistance from some of the middle managers of the company, the first phase of the implementation was working successfully. This implementation is similar to the one depicted in Figure 4.14.

The ERP used Microsoft Access2000™ as the DBMS and a local area network. The system is now able to process 1000 invoices in 20 seconds and is able to store more than fifty million items of data. After the implementation of the system, one of the clerks was made redundant by the company because her work was no longer needed, in spite of the growth of the company's turnover, but the computer technician remained as a part-time employee.

References

Ashcraft, H.W. (2007). *Building Information Modelling. A Framework for Collaboration.* Forum on the Construction Industry. American Bar Association, October 25–26, Newport (Rhode Island).

Björk, B.C. (1999). Information technology in construction: domain definition and research issues. *International Journal of Computer Integrated Design and Construction,* 1(1), 1–16.

Bozdoc, M. (2012). *The History of CAD.* MB Solutions SRL, available 07/12/2012 at http://mbinfo.mbdesign.net/CAD-History.htm.

Chen, Y. and Kamara, J. (2008). The mechanisms of information communication on construction sites. *FORUM Ejournal,* 8, 1–32.

Civitello, A. and Levy, S. (2005). *Construction Operations Manual of Policies and Procedures.* McGraw-Hill, New York.

Date, C.J. (2003). *An Introduction to Database Systems* (8th Edition). Addison Wesley, Reading (MA).

Eastman, C., Teicholz, P., Sacks, R. and Liston, K. (2011). *BIM Handbook: A Guide to Building Information Modeling for Owners, Managers, Designers, Engineers and Contractors* (2nd Edition). Wiley, Hoboken (NJ).

Fisk, E.R. and Reynolds, W.D. (2009). *Construction Project Administration* (9th Edition). Pearson, Upper Saddle River (New Jersey).

Froese, T.M. (2010). The impact of emerging information technology on project management for construction. *Automation in Construction,* 19(5), 531–553.

Harris, F., McCaffer, R. and Edum-Fotwe, F. (2006). *Modern Construction Management* (6th Edition). Blackwell, Oxford.

Hashash, Y.M.A. and Finno, R.J. (2008). Development of new integrated tools for predicting, monitoring, and controlling ground movements due to excavations. *Practice Periodical on Structural Design and Construction*, 13(1), 4–10.

Jung, Y. and Joo, M. (2009). Building information modelling (BIM) framework for practical implementation. *Automation in Construction*, 20, 126–133.

Kimoto, K., Endo, K., Iwashita, S. and Fujiwara, M. (2005). The application of PDA as mobile computing system on construction management. *Automation in Construction*, 14, 500–511.

Lee, G., Sacks, R. and Eastman, C.M. (2006). Specifying parametric building object behavior (BOB) for a building information modeling system. *Automation in Construction*, 15(6), 758–776.

Meena, R.S. (2009). Environmental informatics. A solution for long term environmental research. *Global Journal of Enterprise Information System*, 1(2), 63–69.

Mincks, W.R. and Johnston, H. (2010). *Construction Jobsite Management* (3rd Edition). Thomson, New York.

Onyegiri, I., Nwachukwu, C.C. and Jamike, O. (2011). Information and communication technology in the construction industry. *American Journal of Scientific and Industrial Research*, 2(3), 461–468.

Pellicer, E., Pellicer, T.M. and Catalá, J. (2009). An integrated control system for SMEs in the construction industry. *Revista de la Construcción*, 8(2), 4–17.

Pellicer, E. and Yepes, V. (2007). Gestión de recursos. In: *Organización y Gestión de Proyectos y Obras* (Martinez, G. and Pellicer, E. Eds.), 13–44, McGraw-Hill, Madrid (in Spanish).

Popper, K. (1972). *Objective Knowledge: An Evolutionary Approach*. Clarendon Press. Oxford.

Shi, J.J. and Halpin, D.W. (2003). Enterprise resource planning for construction business management. *Journal of Construction Engineering and Management*, 129(2), 214–21.

Tushman, M.L. and Nadler, D.A. (1978). Information processing as an integrating concept in organizational design. *Academy of Management Review*, 3(3), 613–624.

Vitkauskaite, E. and Gatautis, R. (2008). E-procurement perspectives in construction sector SMEs. *Journal of Civil Engineering and Management*, 14(4), 287–294.

Further reading

Eastman, C., Teicholz, P., Sacks, R. and Liston, K. (2011). *BIM Handbook: A Guide to Building Information Modeling for Owners, Managers, Designers, Engineers and Contractors* (2nd Edition). Wiley, Hoboken (NJ).

Shen, G., Brandon, P. and Baldwin A. (2009). *Collaborative Construction Information Management*. Routledge, London.

5 Site Setup and Construction Processes

5.1 Educational outcomes

The general organisation of a construction site must meet the requirements of the construction project, determining the construction methods, labour force, machinery, equipment, auxiliary and material resource requirements, degree of subcontracting and intermediate and final completion deadlines. By the end of this chapter readers will be able to:

- determine the factors affecting the space available
- understand the equipment and machinery restrictions as well as material storage requirements
- describe temporary facilities and auxiliary works
- define jobsite offices, service facilities and security at the site
- recognise the importance of internal organisation of the construction project.

5.2 Site constraints

Construction site layout is an important planning activity that is solved daily on each construction site. Each construction project needs enough room for temporary facilities such as job offices, material handling equipment, access roads, parking lots, maintenance areas, warehouses, workshops, lay down area for materials, and batch plants to execute the construction works. The space available and its boundaries as well as the requirements of the construction site are essential for the implementation of temporary facilities and the correct

Construction Management, First Edition. Eugenio Pellicer, Víctor Yepes, José C. Teixeira, Helder P. Moura and Joaquín Catalá.
© 2014 John Wiley & Sons, Ltd. Published 2014 by John Wiley & Sons, Ltd.

planning of tasks. Illingworth (2000) indicates three basic site types which have an influence on the construction planning: open field or extensive sites (housing and industrial developments and many civil engineering projects), linear sites (highways, railways and pipelines) and singular or restricted sites (building rehabilitations or industrial improvements). Mawdesley et al. (2002) provide a good discussion of the factors considered during site layout and planning.

Therefore, a good layout must consider the following aspects regarding the plot (Mincks and Johnston, 2010): geographical location, geological conditions, geometry, scale, site topography, ways and paths, location adjacent to the population, current town planning regulations, land qualification, urban equipment, land acquisition, services required and so on. A geotechnical study is also required. In addition, the weather factors must be known with precision (temperature, rain, wind, etc.). Good planning can minimise the damage and disruption of heavy rain to a jobsite. For maritime works, bathymetric studies, the maritime climate, tides and so on are also required.

The correct development of the construction works requires the land for the occupation, but also there must be enough space for the auxiliary facilities and supply of materials, as well as the unavoidable provisional works such as bypasses or cofferdams. Access to quarries or dump sites is another vital requirement. Slopes can ease the circulation of materials because of the gravity effect. In addition, the need to set up a wall or fence must be considered, including for access control. Rainwater can hinder the normal development of the construction project, so that rainwater circulation and evacuation systems can be installed. Works paths and roads should be adequately drained.

As stated previously, construction projects can be classified in three categories (Illingworth, 2000): singular, linear or extensive. Singular construction projects include buildings in undersized plots of land, with the associated problems related to storage of materials, temporary facilities and so on. Linear construction projects (roads, channels or railways) and extensive construction projects (airports, harbours or urban developments) have other problems, such as the transport of materials and equipment around the construction site, reinstatement of services or the access control, fences and security at the site. Figure 5.1 is an example of a site setup for a singular construction project (building). Figure 5.2 illustrates an example of a site setup for a linear or extensive construction project.

The space and the temporary facilities required to execute a construction project are a problem that should be analysed in depth (Illingworth, 2000). Inappropriate layout can result in loss of productivity owing to unnecessary travel time for workers and equipment, or inefficiencies due to safety concerns. Each project is different and requires a slightly different approach. Sometimes both space and location are fixed data of the problem, which means that there is no alternative choice. For example, a building project may need a piece of ground next to the work site. Therefore, whenever possible, it

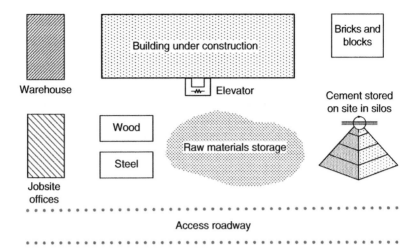

Figure 5.1 Example of the layout of jobsite offices and auxiliary facilities: singular project.

is necessary to devote the necessary time to locate the best option. Note that workers and materials have to be moved from one location to another. Good jobsite location can minimise the movement, which has an impact on the costs of the execution of the work.

Sketches, marked-up blueprints and cut-out templates are frequently used to study space needs and assembly sequences (Henderson, 1976). Even though computer-based tools exist (Rad, 1982; Tommelein et al., 1992a, b), they are seldom used in this field. Nowadays, event simulation based on building information modelling (BIM), explained in Chapter 4, can be used to support construction management. A recent review of the studies that employ these tools for project and construction management can be found in Liao et al. (2011).

Several simple techniques such as flow diagrams, process charts and string diagrams are available for analysing and designing work sites (Harris et al., 2006). The flow diagram indicates the flow of activities on a plan of the construction project. A set of symbols (usually ASME symbols – American Society of Mechanicals Engineers) records the sequence of events and provides a visual image of the processes under study. Sometimes, it is more appropriate to represent the sequence of activities by symbols only. This is the case of the process chart, where the symbols come from the flow chart, but a plan of the site is not required. The use of string diagrams is appropriate when the location of materials and the layout of temporary roads, access points and so on, often develop in a random manner as work proceeds, and this frequently leads to delays and wastage. The string diagram consists of a map of the work area, with the actual movements drawn on top. The objective is to minimise the movements of the worker, machine, material and so on.

5.3 Equipment constraints

The design of facilities at the construction site requires the prior study of the space available and the machinery, equipment and facilities constraints, in accordance with the construction schedule. The equipment required for the construction project has a direct impact on the schedule and the site layout. Location of delivery points, machinery, aggregate, concrete or asphaltic manufacturing plants, steel fixing or carpentry workshops, silos and warehouses, prefabrication plants, compressed air production, generator sets, pumps or other essential parts of the equipment and auxiliary resources at the construction site require their own space and allocation at the site to maximise the efficiency and efficacy of the tasks. Note that crawler cranes and some large mobiles require a fairly large area for rigging. Craneage for the erection of tower cranes or batching plants often requires considerable space (Elbeltagi et al., 2004). In any case, their location must prevent subsequent changes, whenever possible. When this is not possible the changes in location of equipment should not hinder the performance efficiency of the construction project.

Large enough space for all delivery vehicles is required and movements within the site must also be minimised. Machinery, equipment and materials must be adequately transported through the interior using circulation paths, enabling the crossing of any vehicle or establishing marked bypass lanes, with the turning radii and slopes adapted to the type of machinery. The cost of constructing and maintaining haul roads on the site must be considered. These must be in good condition and allow circulation under any climatic conditions, and be supplied with artificial lighting if necessary. It is important to note that dust, mud, frozen ground and snow are factors that impede normal trucks from using haul roads. It is also necessary to check whether there are any obstructions, low bridges, overhead power cables or other structures might prevent the movement of vehicles. Similarly, alternative transport systems must be studied (conveyor belts, cranes, compressed air, fixed cranes, etc.), depending on the type of works and their characteristics. To reduce the cost of such movements, planners must develop a strategy that minimises the total distance travelled by vehicles. In fact, vehicle routing and scheduling is a serious problem that affects the economic results of most distribution and transportation systems (Yepes and Medina, 2006). For example, the shortest route cut and fill problem (SRCFP) seeks to find a route that minimises the total distance travelled by a single earthmoving vehicle between cut and fill locations (Henderson et al., 2003; Lim et al., 2005).

The existing communications network, traffic, link possibilities with facilities, access track or road will determine the decision of entrance and exit points to the construction site. Construction activities should not significantly interfere with existing traffic flow meaning that well-planned temporary traffic management schemes are required. In addition, these inner communication paths must enable the economic traffic of persons, materials, equipment and machinery. In most sites, trucks have to be weighed, which requires the use of the site's own balances, located near the entrance, to enable the easy control of supplies by weight. An appropriate queuing area for waiting trucks may be necessary to avoid congestion of traffic, especially in street areas. It is also convenient to install a petrol station or fuel storage when the dimensions of the construction site allow such a facility.

5.4   Material storage and handling

Material management involves storage, identification, retrieval, transport and construction methods. The efficient handling of materials at the construction site is very important to optimise the productivity of tasks and project profitability. Material should ideally arrive at the works immediately before its use, although this rarely occurs. How materials are delivered and dispatched determines how easily things flow. Thomas et al. (2005) highlight that deficiencies in the supply and flow of construction material will result in causing serious degradations of performance and labour productivity as well as financial losses.

Materials brought onto a construction site are often not used immediately. Some of these materials are valuable and have to be protected from theft; others are dangerous and people have to be protected from them. In any case, they must be stored under a roof or in stockpiles, in order to create a secure reserve to avoid the interruption of the tasks at the site. The most common warehouses at the construction site are those related to cement, aggregates, fuels and lubricants, as well as tools and auxiliary resource warehouses. It is important to estimate the amount of storage needed. Planning and executing material procurement and storage on construction sites is a crucial task to avoid the negative impacts of material shortage or excessive material inventory on-site. The lack of materials can be due to a slower supply rhythm, i.e., supplies are consumed before new stocks arrive, or due to other circumstances, such as delays in transportation, weather conditions or incorrect scheduling.

First of all, the volume of storage required is determined by the cumulative supply and consumption curves. The lower the supply, the greater the anticipation required before the start of the construction works and the greater the volume stored. However, when supplies are not guaranteed, the works should not start until all materials are gathered. Then, when avoiding supply shortages, the supply volume will depend on the probability of interruption and its importance, taking into account economic and delivery deadline effects.

Nevertheless, the storage of materials must be reduced to the bare minimum, since it has an effect on the cost of the facilities required, as well as on loading and unloading procedures, transportation, general costs and surveillance, as well as any possible damage or deterioration caused to the materials. Just-in-time material deliveries are preferred in this case, but they require more coordination with the supplier (Polat and Arditi, 2005). The critical and mutual

> **Box 5.1   Good practices in relation to the handling and stocks of materials**
>
> - minimise the number of movements of materials at the construction site
> - store the materials as close as possible to the place where the tasks have to be carried out
> - place the materials in their storage area
> - use specialised personnel to transport the materials
> - coordinate shop fabrication drawings with the supplier
> - sort the material upon its delivery to the construction site
> - provide appropriate space for sorting and storage waste material
> - do not store materials next to the building as these can obstruct access routes
> - keep all storage areas tidy
> - estimate the storage volume required
> - ensure that there is adequate access to warehouses and stocks
> - compact the storage surface to guarantee the support capacity
> - keep the material on pallets or packaged to avoid damp
> - protect sensitive material from weather effects
> - designate special storage areas for flammable substances such as foam plastics, flammable liquids or gases.

interdependencies between material procurement planning and dynamic material storage and site-layout planning lead to a non-trivial problem that can be solved by heuristic optimisation (Said and El-Rayes, 2011). Box 5.1 (adapted from Thomas et al., 2005, and Minchs and Johnston, 2010) shows some good practices related to the handling and stocks of materials.

5.5   Temporary facilities and auxiliary works

Since the facilities required to execute the construction project, such as electrical power, drinking water or heating, are not available until the works are almost complete, temporary facilities are required to meet the operating requirements during the construction project (Doran, 2004). The contractor must request provisional supply connections for the construction site for each service until the infrastructure is transferred to the owner. Where the company provides only part of the service, it is important to provide the remainder with compatible materials and equipment. Temporary electric power service must be provided as a weatherproof, grounded electric power service and distribution system of sufficient size, capacity and power characteristics during the construction period.

The first task involves establishing connections to the different external supply networks. These points are usually established by the network owners: water, electrical power, telephone network, sewerage and so on. The connections and channels to the site, either overhead or underground, must be planned before foundations or slabs are laid. Temporary utilities include, but are not limited to: water service and distribution; temporary electric power and light; temporary heat; ventilation; telephone service; sanitary facilities, including drinking water; and storm and sanitary sewerage.

When there is no water supply with enough flow near to the construction site, the site's own supply system is installed. In this case, it is important to sterilise temporary water piping prior to use. The terrain's topography will be used with the highest points in the site used for the water supply network. As regards residual water, these should be connected to the existing networks, otherwise they must be adequately treated and drained.

Provisional facilities or auxiliary works also include those required for the execution of the main construction works (Illingworth, 1987). Therefore, works paths, provisional bridges, access slopes to specific areas of the site, bypasses, cofferdams, auxiliary foundations for cranes or scaffolding, quarries with installations for the manufacture of aggregates and sometimes even accommodation for workers and their families during the construction of major works (for example, a dam) all represent temporary facilities or auxiliary works that must be planned before the start of the works and which must be installed and then disassembled after the completion of the works. Another important task is to construct and maintain temporary roads and paving to support the indicated load adequately and to withstand exposure to traffic during the construction period. It is also necessary to extend temporary paving in and around the construction site to accommodate delivery and storage of materials, equipment usage, administration and supervision.

At other times, provisional facilities or auxiliary works must be undertaken to prevent disruption of the work, although they are not strictly necessary to complete the main works, but are obligatory to maintain the existing services affecting third parties as a result of the construction works (Doran, 2004). In this sense, irrigation channels, access to private plots, noise-proof screens, provisional traffic signage, dewatering facilities drains, and such like are examples of temporary facilities that must be taken into account when planning the construction project. Both the scheduling of the construction works and the associated costs must consider the potential demolition or dismantling of these provisional facilities or services.

5.6   Construction jobsite offices

The contractor installs jobsite offices before the start of the construction project. There are as many types of jobsite offices as there are projects and contractors. However, common types may include existing buildings, modular office units, trailers or site-built jobsite offices. In order to select the best jobsite office, several factors should be taken into account, including cost, space at the jobsite and availability (Mincks and Johnston, 2010).

The jobsite office includes, among others, locations (Mincks and Johnston, 2010) the contractor, communication technology resources, archive, information for employees as well as first-aid information and equipment. In short, this office is where the construction site manager, the superintendent and other field management personnel are located. It is usual to install one or a number of prefabricated offices provisionally, with a meeting room with adequate dimensions and an office for the construction site manager, field engineers, supervisors, foremen and technical support and administration departments. The facilities will hold an authorised copy of the project documents and the contract and a construction diary (analysed in Chapter 4). In addition, these facilities may include the offices for the project manager, as a representative of the owner.

These offices must include all modern ICT resources: telephones, computers, printers, fax, internet and so on. Some rooms are prescribed for occupational health and safety purposes: dining rooms, canteens, bathrooms, changing rooms, medical or first-aid services, first-aid kits, warehousing and so on. The dimensions of these rooms depend on the estimated number of workers, with a good degree of comfort required in terms of habitability, safety, health and so on. The personnel's needs and the environmental requirements must also be taken into account: parking spaces, perimeter fences, drainage, evacuation of residual water and rainwater, water connections, electrical energy, communications and so on (Mincks and Johnston, 2010).

The offices should not be located too close to the site in order to avoid the problems related to noise, circulation of machinery and so on. However, they should also avoid excessive or unnecessary travelling of the personnel. In addition, it is convenient that these installations have a view of the work areas and access to the construction site; in this way, the construction site can be controlled from there.

The contractor can also rent a building near the construction site, such as warehouses, office space for rent or even private homes. These cases should consider the costs of adaptation of these infrastructures for their use as jobsite offices. Another option would be the use of modular prefabricated offices or even trailers, which enable a high flexibility in their assembly and disassembly, as well as the possibility of using them later in other locations (Doran, 2004). Finally, when the construction project is going to be developed for a long period of time, buildings are built as offices, which later can be used for other purposes or even dismantled.

Figures 5.1 and 5.2 show examples of different jobsite office layouts and auxiliary facilities. Figure 5.1 is more suitable for a singular construction project, whereas Figure 5.2 fits better in a linear or extensive project.

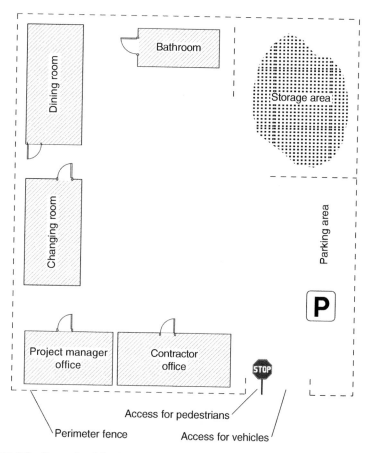

Figure 5.2   Example of the layout of jobsite offices and auxiliary facilities: linear or extensive project.

5.7   Security on construction sites

The construction site must be secured for persons, facilities and equipment. The contractor is responsible for preventing access by the general public, because there are always dangers, such as heavy machinery, excavations, structures, hidden obstacles and so on. These risks are greater for the general public than for the personnel working at the construction site, since the latter are informed, trained and have the necessary experience to reduce the risks. Security and protection facilities include, but are not limited to (Mincks and Johnston, 2010): temporary fire protection; barricades, warning signs and lights; sidewalk bridge or enclosure fence for the site; safety and health and environmental protection.

Another aspect which must be considered is vandalism and theft of materials, tools and equipment from construction sites. Even heavy machinery, such as trucks or bulldozers, have been stolen from many construction sites. This

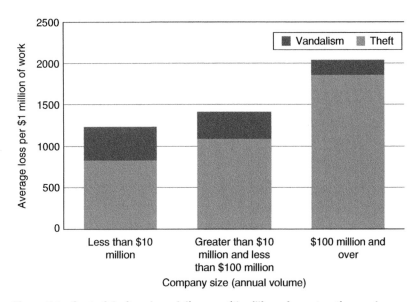

Figure 5.3   Cost of theft and vandalism per $1 million of construction work (primary data obtained from Berg and Hinze, 2005).

problem can affect productivity and diminish the potential profitability of the project being constructed. However, the magnitude of the losses is difficult to estimate accurately. A survey conducted by Berg and Hinze (2005) among 97 firms showed that small to medium-sized companies experienced higher losses from incidents of vandalism when compared to large companies (Figure 5.3). The results of this study show that the annual direct costs of thefts and vandalism incidents were found to total $1682 per $1 million of self-performed work. Thefts account for about 82% of these costs.

Consequently, the contractor must prevent the entrance of unauthorised personnel to the construction site. In addition, the necessary measures must be taken to prevent materials from being stolen (Berg and Hinze, 2005): perimeter fence, security surveillance, controlled access gates, night lighting, communication with local authorities, equipment stored in safe places and so on. Box 5.2 (adapted from Berg and Hinze, 2005) shows some common security practices employed by contractors on construction sites.

Regarding the perimeter fence, in many locations fences can be rented, which includes their transport, installation and disassembly. In other cases, fences can be permanent when a greater security is required, although they are more expensive to disassemble. In any case, it is important to consider the location of access gates and their size to avoid problems with vehicles coming in or going out of the construction site (Doran, 2004). Some additional security measures might require the presence of a guard during the night, usually with a guard dog or a safety patrol. In both cases, a telephone or other method of communication is required to contact the authorities or emergency service if necessary. Other measures include alarm systems, security cameras and security warehouses.

Box 5.2   Most common security practices employed on construction sites

Ensure jobsite security	Prevent theft of tools	Prevent theft of machinery and equipment
Lockboxes	Maintaining a secure storage area	Parking of equipment and machinery in well-lit areas
Security fencing	Marking of tools	Parking of equipment in a specific formation at the end of the day
Posting of warning signs	Maintaining tool inventory	Including additional identification on equipment and machinery
Use of exterior lighting	Minimising tools left on site	Using a distinctive colour to mark machinery and equipment
Strategic parking of large equipment	Making workers responsible for tools	Modifying the ignition or fuel lines

5.8   Internal organisation of the construction works

The construction project is organised in a hierarchical structure, subject to a series of rules and codes of behaviour that enable the construction company to reach the economic, time, quality and safety objectives with efficacy and efficiency (Mustapha and Naoum, 1998). In order to reach these objectives with a coordinated approach, the activities are grouped in departments or sections, with a clear distribution of functions and responsibilities, where each person knows the role they must fulfil and the way in which their tasks are related to others.

The internal organisation of the construction project has functional levels established in an organisational chart (Mustapha and Naoum, 1998). This organisational chart displays the interrelation standards among posts or positions, defined by a series of norms, directives or internal regulations required to attain its objectives. Each construction company has a different way of organising the construction project, adapting their operations to the particular characteristics of each construction site.

Appropriate organisation of the works has economic, time, safety and quality advantages. Nevertheless, there are no two identical projects, since each of them is developed on a different site, with many temporary employees. Also, the projects experience different changes throughout the construction phase due to unforeseen circumstances, deficiencies or other types of events. Therefore, the organisation of the construction site is decisive for the project's success.

The dynamics of the construction project make a regulated organisation non-viable, if it is based on rigid and pre-established norms. A structure with these characteristics, typical of public administrations, has the advantage of solving similar problems with the same approach, but production is generally slow and bureaucratic, that is, there is not enough flexibility to adapt to the changing situations present at each construction site.

The construction project is usually organised with a linear approach (Halpin and Woodhead, 1998). The structure is simpler and is characterised by the linear authority principle, where the communications among members of the organisation follow the hierarchical line established and the transmission of orders, tasks and responsibilities is clear and precise. The advantages of a linear organisation include a clear and stable unit, with a simple implementation. This type of organisation fits well especially in small and medium-size construction projects, which are not too specialised and have standardised and routine tasks with defined execution procedures.

In other construction projects, owing to their dimensions, complexity or long execution times, there is a need for consultants or specialised departments. This is the case of a functional organisation, with a management style based on knowledge, whereby none of the supervisors or managers can exert full authority over all employees. This type of organisation facilitates the decentralisation of decisions and direct communication, with no intermediaries (Robbins and Coulter, 2010). However, construction sites with a purely functional organisation could lead to the loss of control authority, a multiple subordination of different specialised departments and probably confused objectives.

To avoid these problems and increase their organisational advantages, construction projects with a determined complexity propose a hierarchical-advisory organisation (Reve and Levitt, 1984). In these types of organisations, the sole authority principle is maintained and the advisory or support bodies help their line managers to make decisions. The hierarchy (line) guarantees control and discipline, while specialists provide the consulting and advisory services.

Medium-size construction projects are usually managed by the construction site manager, with three departments usually reporting to this position: the technical support, operational or productive (construction per se) and administration. Figure 5.4 shows a standard organisation chart.

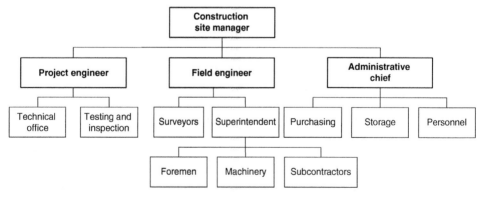

Figure 5.4 Organisational structure of the construction site.

The construction site manager assumes the responsibility and the objectives assigned to the construction project. They usually report to the portfolio manager or to the director of construction operations at the construction company (see Chapter 1). The construction site manager is responsible for the following tasks (Edum-Fotwe and McCaffer, 2000):

- represents the company in front of the client, the subcontractors, and any other third parties
- leads the resources at the site, including the personnel
- clarifies any issues regarding the definition of the project, together with the project manager representative of the owner
- drafts the bill of quantities for work units, materials, subcontractors and so on
- decides about the layout of plants, stocks, warehouses and workshops
- analyses the construction processes
- schedules the construction tasks
- coordinates and monitors the execution of the construction project
- cooperates with the project manager in the drafting of valuations and final payment.

The construction site manager is also responsible for administrative management tasks: reception and storage of materials, consumption of materials, works inventories, contracting and management of personnel, assessment of subcontractor activities, management of machinery and consumption of fuel, electricity and so on. As regards the execution of the construction project, the construction site manager leads the preparatory operations before the start of the construction works, ordering the personnel's activities and machinery operations, as well as the implementation of materials (Mincks and Johnston, 2010). They also have responsibility for the control of the site and the subcontractors, work reports and occupational health and safety.

Nevertheless, the construction site manager must have an adequate organisation to attain these objectives (Halpin and Woodhead, 1998). The field engineers, superintendent and foremen complete the organisation to develop the construction project. One of the most important figures is the superintendent, who should have a vast experience in the organisation, management and supervision of construction activities and processes with a direct and close approach. The superintendent is the hierarchical link between the workers and the management team. In important projects, with different sites or specialised units, there can be various superintendents coordinated by a general superintendent.

Foremen are the vehicle that links superintendent with the operators assigned to a site (Mincks and Johnston, 2010). They are usually recruited because of their vast experience and responsibility. They look after the performance of the workers, are in charge of meeting deadlines and the order of workers, as well as their training, whenever required. They fill in the daily manpower, equipment and machinery and material reports. In addition, they propose all appropriate changes, modifications and controls to the management areas.

The technical support services are not directly on the production line but rather report to the construction site manager. This department is in charge of:

- technical office: detailed design, studies, calculations, valuations and cost control
- quality and environmental management: laboratories and technical control.

The administrative services also report directly to the construction site manager, out of the executive line. They are responsible for:

- material and tool purchasing orders, storage, distribution and control
- management of facilities and equipment: workshops, maintenance, machinery plant and so on
- administrative and legal matters related to the personnel
- registry of accounting operations
- collections and payments
- other support tasks: correspondence, typing, filing and so on.

5.9   General approach to construction processes

The construction process comprises a set of activities, methods of organisation and arrangement of construction elements and systems that allow a project to be executed using the labour force, machinery and material resources available (Illingworth, 2000). The process includes the design, execution, operation and maintenance and, at times, site dismantlement and closure. It is important to understand the distinction between construction processes and procedures in building and those in civil engineering.

Building needs a series of techniques, called construction procedures, which are used to execute the activities of a building project. Many of the techniques used in building are found in civil engineering projects too. However, building also involves specific techniques for certain constructions such as partition walls, roofs or installations. Therefore, traditional construction processes associated with foundation laying activities (screen walls, pile driving and so on) must also include tasks which are vital to construction works such as the formwork and the removing of the formwork. The latter clearly affects both deadlines and the loads on each slab, which means that the formwork has to be reapplied at times.

Industrial buildings are also included in this field. These buildings encompass industrial production operations, raw material deposits and warehouses, engine rooms, chimneys or structures dedicated to transportation or extraction. In addition, major construction projects (sports arenas, exhibition centres, etc.), as well as very tall buildings, require specific construction methods and procedures; new light materials and the industrialisation and prefabrication of work execution play an increasingly important role (Van Bueren and De Jong, 2007).

The construction of a building's systems should comply with all regulatory requirements, and contribute to the sustainability of the building (Van Bueren and De Jong, 2007). This can be achieved by using traditional systems and procedures, which are not usually very efficient as they include an excessive use of manpower and manual construction techniques. Alternatively, more rational and industrialised techniques can be used, in which prefabrication is becoming increasingly important.

The way in which construction elements and systems in civil engineering are organised and arranged, and the set of techniques required to build the different units, depends mainly on the type of construction project (Illingworth, 2000). Therefore, the determining factors of a harbour are different from those of an underground project (a tunnel) or a structure (a bridge). Nevertheless, a general description of the most common construction procedures in this field is worth considering.

Drilling and boring are essential to analyse the land in any type of project, but the use of soil improvement techniques is common (preloads, grouting, stabilisations, etc.) to increase the support capacity, limit deformations or control the groundwater level. The next step consists of earthworks or dredging, which are the most important processes in some works (roads, restoration of beaches, etc.). Occasionally rocks must be blasted, either on the site itself, or in the quarry that supplies materials to the construction works. In some cases, blasting procedures are employed in the construction of tunnels and underground works, although specific machinery can also be used, such as tunnelling machines or shields, as well as other techniques, such as the Austrian method (Rabcewicz, 1964). After the earthwork activities, the soil is compacted. This process is particularly important in civil works such as the execution of earth and rock-fill dams. It is also fundamental in linear construction works, such as roads, where asphalt or concrete road execution procedures are required.

The execution of structures can involves surface and deep foundation laying activities (piles, sheet piles, sheet walls, caissons, etc.) and might need auxiliary structures, such as falsework and formwork, to contain and pour concrete (Harris et al., 2006). In other procedures, prefabrication plays a key role, such as the construction of bridges with precast sections.

Finally, the dismantlement and disassembly of provisional or even final building works should be considered. These activities could include the use of explosives for demolition, and also special works such as the landscaping of an abandoned quarry or the construction of a concrete caisson for the reactor of a nuclear power station (Seike and Akita, 2008).

5.10   Temporary works

The expression 'temporary works' is widely used in the construction industry and covers any 'parts of the work that allows or enables construction of, protect, support or provide access to the permanent works, and which might or might not remain in place at the completion of the works' (BSI, 2008). It is frequently necessary to design, execute and maintain supplementary construction works during the execution of the main construction works as they are necessary for its completion. For example, on a road, traffic detours must be built to ensure that the traffic can flow freely during the construction of a bridge. Sometimes a provisional foundation slab is required to support a crane in a port during the construction of a breakwater. Another example is the execution of a sheet pile with metallic profiles, which facilitates soil containment during earthwork movements. Temporary works also includes the activities needed to stabilise or protect an existing building or structure that are not intended to be permanent.

Temporary works are likely to form a key part of any construction process (Harris et al., 2006). For example, the preload technique improves the bearing capacity of soil and also decreases its deformability. This is achieved through the accumulation of material over a long period of time (months) and its subsequent removal. Another example of temporary works is the execution of provisional structures, such as the formwork and shuttering needed to pour concrete for a bridge built on site. In addition, the correct design and execution of temporary works is an essential element of risk prevention and mitigation in construction. But poorly planned, designed, constructed, supervised and managed temporary works can leave projects open to risks and the consequent delays and increased costs.

In some cases, the aim of the temporary works is to maintain the services or supplies affected by the main works (Illingworth, 1987). Sections of channels or irrigation ditches, access roads, provisional drains or noise-proof screens are some of the works or facilities that are implemented for the duration of the project. Examples of temporary works include:

- earthworks: trenches, excavations, temporary slopes and stockpiles
- structures: formwork, falsework, scaffolding, temporary bridges, propping, facade retention, needling, shoring, edge protection, site hoarding and signage, site fencing, cofferdams
- equipment/plant foundations: tower crane bases, supports, anchors and ties for construction hoists and mast climbing work platforms, ground works to provide suitable locations for plant erection.

Although this type of construction work is temporary, it requires suitable planning at the design, execution and dismantlement phases. A common practice is to simply rely on the experience gained in similar cases, but this can lead to accidents or unacceptable risks during construction. Therefore, unless these construction works are defined in the main design project, they require both a specific design project signed by a certified technician and site supervision to guarantee the safety of people and goods while the main construction activities are being carried out (Illingworth, 1987).

It is also necessary to establish in advance how the temporary works will be demolished, dismantled or disassembled, or how the areas and services affected by the construction works will be restored. So it should be remembered that using recyclable or modular materials that can be disassembled can lower execution costs, and reduce the environmental impacts of this type of temporary construction works and facilities. If temporary construction works are not dismantled or disassembled after completion of the construction project, they should receive an official license and, where necessary, their activities should be recalculated to ensure they have a longer useful life than originally scheduled (Seike and Akita, 2008). In fact, many standards prescribe lower partial safety coefficients than those required for final construction works. In addition, the durability factors of materials should be reviewed (metallic oxidation, insufficient coating on concrete frameworks, etc.) and also factors associated with functionality (unacceptable deflection, appearance of cracks, etc.).

Box 5.3 indicates the main issues to be addressed in temporary works following the procedure established by the British standard BS 5975 (BSI, 2008).

Box 5.3   Items to be addressed in temporary works (according to clause 6.3.1 of BS 5975)

- appointment of a temporary work coordinator for health and safety matters
- on larger sites, appointment of temporary site supervisors
- assessing/ensuring competence of design and site-based staff responsible for temporary works
- preparation of adequate design briefs
- designing to include calculations, sketches, drawing, specifications and design risk assessments
- independent checking of the design
- issuing of design checking certificates
- procurement of temporary works materials and equipment (in accordance with the designer's specifications)
- site control of erection, use, maintenance and dismantling of temporary works
- checking of erected temporary works (and control of their use) to ensure compliance with the design
- issuing of the 'permit to load' and 'permit to dismantle' when required.

References

Berg, R. and Hinze, J. (2005). Theft and vandalism on construction sites. *Journal of Construction Engineering and Management*, 131(7), 826–833.

BSI (2008). *BS 5975: Code of Practice for Temporary Works Procedures and the Permissible Stress Design of Falsework*. British Standards Institution, London.

Doran, D. (2004). *Site Engineers Manual*. Whittles Pub., Dunbeath, United Kingdom.

Edum-Fotwe, F.T. and McCaffer, R. (2000). Developing project management competency: perspectives from the construction industry. *International Journal of Project Management*, 18(2), 111–124.

Elbeltagi, E., Hegazy, T. and Eldosouky, A. (2004). Dynamic layout of construction temporary facilities considering safety. *Journal of Construction Engineering and Management*, 130(4), 534–541.

Halpin, D.W. and Woodhead, R.W. (1998). *Construction Management* (2nd Edition). Wiley, New York.

Harris, F., McCaffer, R. and Edum-Fotwe, F. (2006). *Modern Construction Management* (6th Edition). Blackwell, Oxford.

Henderson, D., Vaughan, D.E., Jacobson, S.H., Wakefield, R.R. and Sewell, E.C. (2003). Solving the shortest route cut and fill problem using simulated annealing. *European Journal of Operational Research*, 145(1), 72–84.

Henderson, E.M. (1976). *The Use of Scale Models in Construction Management*. Technical Report No. 213, Department of Civil Engineering, Stanford University, Stanford (California).

Illingworth, J.R. (1987). *Temporary Works – Their Role in Construction*. Thomas Telford, London.

Illingworth, J.R. (2000). *Construction Methods and Planning* (2nd Edition). E & FN Spon, London.

Liao, T.W., Egbelu, P.J., Sarker, B.R. and Leu, S.S. (2011). Metaheuristics for project and construction management: a state-of-the-art review. *Automation in Construction*, 20(5), 491–505.

Lim, A., Rodrigues, B. and Zhang, J. (2005). Tabu search embedded simulated annealing for the shortest route cut and fill problem. *Journal of the Operational Research Society*, 56(7), 816–824.

Mawdesley, M. J., Al-Jibouri, S. H. and Yang, H. (2002). Genetic algorithms for construction site layout in project planning. *Journal of Construction Engineering and Management*, 128(5), 418–426.

Mincks, W.R. and Johnston, H. (2010). *Construction Jobsite Management* (3rd Edition). Thomson, New York.

Mustapha, F.H. and Naoum, S. (1998). Factors influencing the effectiveness of construction site managers. *International Journal of Project Management*, 16(1), 1–8.

Polat, G. and Arditi, D. (2005). The JIT materials management system in developing countries. *Construction Management and Economics*, 23(7), 697–712.

Rabcewicz, L. (1964). The New Austrian Tunneling method: part one. *Water Power*, 15(11), 453–457.

Rad, P.F. (1982). A graphic approach to construction job-site planning. *Cost Engineering*, 24(4), 211–217.

Robbins, S.P. and Coulter, M. (2010). *Management* (11th Edition). Pearson, New York.

Reve, T. and Levitt, R.E. (1984). Organization and governance in construction. *International Journal of Project Management*, 2(1), 17–25.

Said, H. and El-Rayes, K. (2011). Optimizing material procurement and storage on construction sites. *Journal of Construction Engineering and Management*, 137(6), 421–431.

Seike, T. and Akita, N. (2008). Dismantlement of buildings and recycling of construction waste. In: Hanaki, K. (ed.) *Urban Environmental Management and Technology*, 107–128. Springer, Tokyo, Japan.

Thomas, H.R., Riley, D.R. and Messner, J.I. (2005). Fundamental principles of site materials management. *Journal of Construction Engineering and Management*, 131(7), 808–815.

Tommelein, I., Levitt, R. and Hayes-Roth, B. (1992a). Sight plan model for site layout. *Journal of Construction Engineering and Management*, 118(4), 749–766.

Tommelein, I., Levitt, R. and Hayes-Roth, B. (1992b). Site-layout modeling: how can artificial intelligence help? *Journal of Construction Engineering and Management*, 118(3), 594–611.

Van Bueren, E. and De Jong, J. (2007). Establishing sustainability: policy successes and failures. *Building Research & Information*, 35(5), 543–556.

Yepes, V. and Medina, J.R. (2006). Economic heuristic optimization for heterogeneous fleet VRPHESTW. *Journal of Transportation Engineering*, 132(4), 303–311.

Further reading

Doran, D. (2004). *Site Engineers Manual*. Whittles Pub., Dunbeath, United Kingdom.

Illingworth, J.R. (2000). *Construction Methods and Planning* (2nd Edition). E & FN Spon, London.

6 Machinery and Equipment

6.1 Educational outcomes

The machinery and equipment required to execute the activities at the construction site have a direct impact on productivity and production costs. They affect the profitability of the project as well as the company. By the end of this chapter, readers will be able to:

- understand the need to use machinery and equipment at the construction site
- describe the main factors and criteria for the selection of machinery and equipment
- determine the fixed and operating costs of machinery and equipment
- recognise the importance of the maintenance operations.

6.2 The need for machinery and equipment

The mechanisation of activities in civil engineering and building is necessary from the technical, economic, human and even legal perspective. The purpose of machines and equipment is to release human beings from the hardest tasks; they have become a tool to produce a greater quantity of product, cheaper and with greater quality. They enable human beings to reduce the time required to carry out tasks, and some would have been impossible to perform in the past. The awarding of a contract usually requires the construction company to provide adequate machinery and equipment in order to guarantee the conditions, quality and safety requirements of the construction project. Also, some activities are not feasible without the use of machinery

Construction Management, First Edition. Eugenio Pellicer, Víctor Yepes, José C. Teixeira, Helder P. Moura and Joaquín Catalá.
© 2014 John Wiley & Sons, Ltd. Published 2014 by John Wiley & Sons, Ltd.

and equipment, such as grouting, pile driving and dredging. In other cases, the manual execution of concreting, land compacting, and such like could not satisfy the level of requirements stated in current specifications.

Machines and equipment constitute a significant portion of the investment in fixed assets by contractors engaging in equipment-intensive projects such as earthmoving, hydraulic infrastructure, highway, maritime works and industrial installations. The total construction and mining equipment replacement value of the top 250 companies in North America had reached nearly US$100 billion in 2006 (Stewart, 2006). The rate of investment in machinery and equipment, calculated in relation to the annual cost of acquisition and the total annual time, varies from 3% to 13%. The mechanisation rate is between 13% and 19% (value of machinery and equipment in relation to the annual production) for construction companies.

6.3   Selection of machinery and equipment

The acquisition of machinery and equipment is an essential aspect of the economic profitability of every construction company. Therefore, the selection of machinery and equipment should be based on a comprehensive economic study. The determining factors and selection criteria are explained in the following paragraphs.

6.3.1   Conditioning factors

The economic and financial situation of a company, the type of works to be carried out, the location of the machinery and equipment, the company policy and strategy and the perspective of new contracts are factors that determine the choice of machinery and potential acquisition. However, the circumstances and typology of tasks, production capacity required, flexibility to the changing conditions, reliability and after-sales service are associated conditions that also determine the choice of the most convenient type of machine.

The selection process is aimed at the unification of equipment: greater simplicity in terms of handling, maintenance and repairs, smaller spare part inventory, simplification in the training activities and documentation. Furthermore, machines are deemed adequate when they are able to work in groups. Other issues to be taken into account are the analysis of their maintenance costs (fuel consumption, conservation materials and parts that can become worn) and the estimation of their forecast production. In addition, the transport costs must be assessed, both ways, including the cost of assembly and disassembly, licence plate registration and insurance. The best choice of equipment for a specific type of construction site will depend on various factors (Day and Benjamin, 1991): geographical location of the site, access, climatology, topography, available energy, term of execution, type and extension of the construction project, time allowed to do the job, balance of interdependent equipment, mobility requirement of the equipment, versatility of the equipment and so on.

Most construction projects have to be carried out with more than one kind of equipment, generally combining different types of equipment. Moreover, there are different brands and models of machines capable of satisfying the requirements of each type of construction project. The best option depends on the price of the machine, forecast performance and factors such as the quality of the after-sales service, price of spare parts and speed of supply. The prices in catalogues are subject to important variations, such as discounts, availability of second-hand (used) machinery, payment conditions and interest levels. The forecast performance is a very important factor in the choice of the equipment, but it is difficult to estimate without having prior experience or real demonstrations of the model. Finally, the acquisition of new machines is as valid as purchasing used machines, or taking other financing schemes, such as leasing. These options involve direct or indirect ownership of the equipment, and special facilities are needed for servicing and maintenance, together with experienced management and skilled staff. Consequently, the alternative of hiring plant should not be ignored (Halpin and Woodhead, 1998). Box 6.1 (adapted from Peurifoy et al., 2010) indicates the differences between a lease and a loan.

6.3.2 Methods used to select the machine in relation to economic profitability

To select the best possible purchase option, a study is usually performed, which maximises the economic profitability, considering the monetary updating of the investment whenever required (Peurifoy et al., 2010). There are two methods with no monetary updating:

- recovery of the investment or return period: The machine that minimises the time required for the future net profits and equals the price of acquisition of the investment is selected. This method does not take into account the profitability of the investment
- average profitability of the investment: The machine that offers the highest average performance rate is selected; in other words, the greatest quotient between the sums of all net profits generated throughout the investment's working life and the cost of acquisition.

Other methods do consider monetary updating (Gransberg et al., 2006). The value of money depends on time, since interest is levied on the availability of the money loaned. Therefore, given an interest rate i in unit terms and n periods of time, a current quantity P and a future quantity S will be related as:

$$S = P \cdot (1+i)^n \tag{6.1}$$

In this case, the inter-temporary comparisons of monetary units must be obtained with the updated revenue or expenses. In addition, these calculations

Box 6.1   Differences between a lease and a loan

Lease	Loan
A lease requires no down payment. It finances only the value of the equipment expected to deplete during the term of the lease. The lessee usually has the option to purchase the equipment at the end of the lease for its remaining value.	A loan usually requires the end user to make a down payment on the equipment. The loan finances the remaining amount.
The leased equipment itself is usually all that is needed to secure the transaction.	A loan often requires the borrower to provide other assets as collateral.
A lease requires only a lease payment at the beginning of the first payment period, which is usually much lower than the down payment.	A loan usually requires two fees during the first payment period: a down payment at the beginning and a loan payment at the end.
The end user may claim the entire lease payment as a tax deduction. The equipment write-off is tied to the lease term, which can be shorter than the depreciation schedule, resulting in large tax deductions each year. The deduction is also the same every year, which simplifies the accounting.	Borrowers may claim tax deduction for a portion of the loan payment as interest and for depreciation, which is tied to the depreciation schedule.
Leased assets are paid when the lease is an operating lease. Such assets do not appear on the balance sheet, which can improve financial ratios.	Financial accounting standards require owned equipment to appear as an asset with a corresponding liability on the balance sheet.
Most of the cash flow, especially the option to purchase the equipment, occurs later in the lease term when inflation makes money cheaper.	A larger portion of the financial obligation is paid at current currency value (more expensive).

must consider the inflation expectations. However, future inflation usually leads to the increase in monetary value, so that the performance and costs would be the same. Therefore, the following must be considered:

- net present value: The machine that maximises the difference between the present value of net revenue and the cost of the investment is selected (NPV), where e_j are the net revenues in year j, n the number of periods and i the interest rate. The present value of revenues is calculated as:

$$NPV = \sum_{j=1}^{n} \frac{e_j}{(1+i)^j} - V_a \qquad (6.2)$$

- internal rate of return: The machine with the highest internal rate of return (IRR) is selected, defined as the value i that cancels the NPV. One

of the advantages is that i is not required for its calculation. The investment will be appreciate if the IRR exceeds the interest rate of the market.

Even though the IRR cannot be considered as an alternative to the NPV, it is
interesting to calculate its value as supplementary information.

6.4   Calculation of costs

Equipment managers must be able to forecast costs accurately. This task not
only helps ensure realistic profits for contractors, but it can also ensure that
the project is delivered within budget for clients. Usually, profit maximisation
or cost minimisation is the main objective of any economic model for equipment management (Douglas, 1975). The remainder of this section focuses on
forecasting equipment and machinery costs.

6.4.1   Fixed and variable hourly costs

The hourly cost of the machinery and equipment is divided in two components (Chen and Keys, 2009):

- fixed or ownership costs (amortisation, investment costs, insurance,
  maintenance and cleaning). They are independent of the number of hours
  of use of the machine
- variable or working costs (repairs, operator's labour, fungible elements,
  fuel, lubricants, transport and installations, as required). They are inherent to the machine's operation.

There are also concepts, such as the labour, that are included in both groups
(the driver in charge of the conservation and maintenance of the unit while it
is at rest). The total cost of the investment C_T (in a unit with a forecast working life of N hours) depends on the fixed unit costs G_f and variable unit costs
G_v. It is calculated according as:

$$C_T = A \cdot G_f + B \cdot (G_f + G_v) \qquad (6.3)$$

A = hours stopped
B = working hours
$N = A + B$ = working life of the machine in hours

The relation between the effective working hours and their stopping times has
an effect on the hourly cost of the equipment. The hourly cost, C_h, is easily
calculated by dividing the total cost by the number of hours of operation of
the machine. Equation 6.4 shows how the machines are working without
stops during the maximum time possible in order to be fully productive.

$$C_h = \frac{C_T}{B} = \frac{N}{B} \cdot G_f + G_v \qquad (6.4)$$

6.4.2   Equipment ownership costs

Amortisation

The amortisation of the machine is the monetary quantification of its depreciation, which is defined as the decrease in the residual value of equipment over time because of consumption, wear and tear or obsolescence. Cross and Perry (1995) conducted one of the most comprehensive studies on depreciation of agricultural equipment; this analysis can be translated to construction too. The purpose of an amortisation is for:

- creation of a fund for the renovation of the machine
- reflecting the decrease in patrimonial value of the company in accounting terms
- distribution of the cost of machinery throughout the production
- recovering the purchase (investment) money over several years.

The depreciation value V_d is the quantity considered in the depreciation of the equipment, equivalent to the difference between the purchase value and the residual or resale value. The machine owner, for accounting purposes, estimates the resale value in advance. This residual value is defined as the expected selling price in the market at a point of the equipment's service life. For many owners, the potential resale value is a key factor in their purchasing decisions. Several methods are used, going from simple rule of thumb to statistical regression analysis, in order to assess this value (Lucko et al., 2006). But recent approaches such as data mining demonstrate advantages, as well as increasing accuracy, to predict the residual value of equipment (Fan et al., 2008). Tyres are not amortised with the rest of the equipment since they have a shorter working life than the machine itself.

The initial investment or cost of acquisition V_a of the machinery is the factory price, including loading, transport, packaging, transport insurance, assembly and disassembly, until it is ready to work at the construction site. The cost of tyres is considered separately from the cost of acquisition, valued as a fungible element, because of their shorter working life.

The fluctuating price of the machinery throughout its working life requires the establishment of a representative and invariable price. The average investment value V_m is the quotient between the sum of the equipment's valuation at the start of each year and the working life in years. The interest, insurance, taxes and storage costs are calculated using this value. The amortisation payments are calculated in periods (years or working hours), although they can be assessed in relation to the tasks developed or work units produced. However, the latter has a complex practical application since the production depends on the performance, which varies because of the application of multiple factors, such as the working conditions or organisation of the construction project.

The depreciation period will vary according to the working conditions of the equipment. If the amortisation is a commercial measure aimed at recovering the investment, its period should be the same as the working life of the equipment. This concept is called a technical amortisation. The following factors must be taken into account to establish the number of working life hours (Gransberg et al., 2006):

- use of the machine
- working conditions
- maintenance schedules
- statistical data on equivalent machines
- handling skill of the operator.

Nevertheless, other tax or accounting criteria are used in many situations. Therefore, the company can amortise the machine within a shorter period of time, which will lead to less profits during the first few years since the amortisation is an expense and will be deferred in terms of the tax burden. Therefore, laws and regulations establish the fiscal or accounting amortisation.

There are several methods that can be used to establish the machinery's depreciation in accounting terms (Day and Benjamin, 1991; Halpin and Woodhead, 1998; Gransberg et al., 2006). It is hard to adjust the amortisation to the real loss in value of each unit in time. The most common systems used are proportional or linear systems, as well as variable and degressive amortisation. Degressive methods are closer to reality, since the depreciation is greater at the beginning. Nevertheless, in many cases linear amortisation is used. The **straight-line method** shows the uniform decrease in the cost of acquisition, which enables the distribution of the amortisation payments in years, months or even days; depreciation can be calculated as:

$$D = \frac{V_d}{N} \tag{6.5}$$

D = depreciation
V_d = depreciation value
N = number of working life years

The average value V_m of an investment V is expressed as:

$$V_m = \frac{1}{2N}(N+1) \cdot V \tag{6.6}$$

In equation 6.6 the investment value V can match the depreciation value V_d. For example, the residual value does not have to be amortised and, in turn, it must be included in the calculation of the average value in order to compute the interest and other similar concepts. A variation in this procedure considers the double annual payment of an amortisation during the first year. This is how the residual value can be better calculated to represent the real value of the machine at the end of each year.

In the **sum-of-the-years-digits method**, each year is depreciated by a coefficient; its denominator is the sum of digits of the machine's working life in years, and its numerator is, for the first year, the number of years, for the second year, the number of years minus one, and for the final year, the unit. It is a simple calculation, although fractions of years cannot be used. Following this notation, it is easy to demonstrate (see equation 6.7) how the average value of the investment is the sum of values at the start of the amortisation years divided by the number of years N:

$$Vm = \frac{(N+2)}{3N} \cdot V \tag{6.7}$$

The **declining-balance depreciation method** can be applied when periods of use are to be very short (2–4 years). This involves a depreciation that is double the linear case, applied during the first year. During the second year, the same double application is applied to the difference between the acquisition value and the depreciation of the first year. This is calculated consecutively for the remaining years. In this case, the value is not null during the final year, but it is reasonable to consider it as the residual value. The residual value V_r at year N is formulated as:

$$V_r = V \cdot \left(1 - \frac{2}{N}\right)^N \tag{6.8}$$

Considering that the unit is amortised at the acquisition value, the average value of the investment will be the sum of values at the start of each year of amortisation, divided by the number of years N:

$$Vm = \frac{V}{2}\left[1 - \left(1 - \frac{2}{N}\right)^N\right] \tag{6.9}$$

Other systems used to calculate the depreciation are as follows:

- proportional over the production: the machine is amortised in accordance with the forecast production value instead of considering its working life. This method can be risky since the working conditions at each site are different
- fixed (or American) payment method: consists of determining the amortisation payment (sum of the amortisation plus interests) at a compound interest:

$$a = \frac{V \cdot r}{1 - (1+r)^{-N}} \tag{6.10}$$

a = annual amortisation
V = capital amortised
r = rate by one of the compound annual interest
N = number of working life years

The annual depreciation, when there is a residual value V_r at the end of its working life, is computed as:

$$a = \frac{\left[V - V_r \cdot (1+r)^{-N} \right] \cdot r}{1 - (1+r)^{-N}} \tag{6.11}$$

The fixed payment method is far from reality, since the depreciation is greater during the first years, although it enables the calculation of the hourly cost of the machinery, with a constant cost. It includes not only the amortisation but also the interest.

The declining-balance method and the sum-of-the-years-digits formulas are designed to put most of the depreciation at the beginning of the working life. They provide a quick settlement that goes hand in hand with the real loss of the equipment's value in normal markets. The problem is the calculation of hourly costs, which would be variable, in accordance with the number of years. Therefore, it seems better to use straight-line depreciation for the calculation of hourly costs. With this method, the first annual payment is frequently doubled to compensate for the strong depreciation during the first year. Box 6.2 details an annual depreciation example comparing the three most used methods: straight-line, declining-balance and sum-of-the-years-digits; it includes a display of the results in Figure 6.1.

Indirect costs

The cost of the insurance policy for the machinery and equipment covers the risks (fire, theft, accidents, etc.) whereby the value can be fully or partially lost; this may require an investment to replace the machine under normal operating conditions (Caterpillar, 2011). These annual costs are applied proportionally to the equipment in accordance with the value considered for that year; that is, they are applied to the average value of the investment. An average percentage of 2% is commonly used when no other information is available. The remaining costs will vary depending on the machinery and transport, but the average admissible values are usually 2% for taxes and 1% for storage, also calculated with the average value of the investment and in annual terms.

If the equipment ties up money provided by a financing company, they expect a return. Similarly, if the equipment ties up money provided by owners or stakeholders, they also expect a return. Thus, interest is considered as the cost of using capital. It is usually treated as another fixed cost of the machinery, since some companies prefer to include it in the general costs of total operations. The choice of interest in machinery renewal studies is vital. The interest must not be below that of the money loan, but it also should not be so high as to cause a problem when the equipment needs to be replaced and so delay its replacement.

If the American amortisation method is used, the amortisation payments can be calculated with the sum of the amortisation and interest, with a

Box 6.2   Annual depreciation example

This table gives annual depreciation on €1000 for a period of five years comparing the three most used methods: straight-line, declining-balance and sum-of-the-years-digits. It can be applied for calculations on any price of machine, by multiplying by its cost divided by 1000.

Year	Straight-line	Declining-balance	Sum-of-the-years-digits
1	200.00	400.00	333.33
2	200.00	240.00	266.67
3	200.00	144.00	200.00
4	200.00	86.40	133.33
5	200.00	51.84	66.67

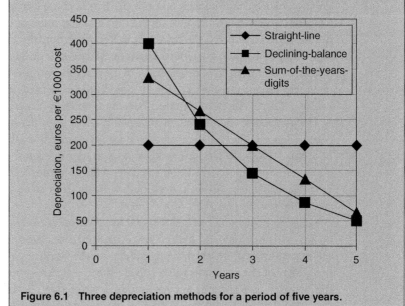

Figure 6.1   Three depreciation methods for a period of five years.

constant result for the calculation of the hourly cost, since this method applies constant annual payments. If linear amortisation method is used, an approximate fixed annual sum for the total interests is obtained, applying the average value interest of the investment. In fact, the amortisation is equal for all annual payments, but the payment of interest varies according to the capital pending amortisation at the start of each year, decreasing these payments every year.

The maintenance costs depend on the type of machine and its working conditions. To calculate the hourly cost, a percentage between 4% and 8% of the hourly linear amortisation cost is applied. If the amortisation is variable, the percentages related to the maintenance costs will also be variable.

6.4.3   Operating costs

Operating costs are the owner's costs to have the equipment operating at the job site. These costs are usually bigger than the ownership costs and they are essential factors related to machine operation. Energy sources and operators' wages are explicit operating costs, which are only considered when the machine is running. On the other hand, implicit operating costs are costs that actually take place when the machine is in downtime (Chen and Keys, 2009).

Lubricants and fuels

Consumption is classified in two groups: main and secondary. Main consumptions correspond to fuels (gas oil, petrol or electrical energy) and secondary consumptions are lubricants, oils, filters, grease, cotton and so on. The cost of fuel will depend on price and consumption. The price depends on factors such as the current oil price per barrel, taxes, transport and so on. Consumption strongly depends on the type and condition of the equipment, the working conditions, the height of the construction site above sea level, temperature, weather, the temperament of the operator or the hourly performance of the machine. For example, two operators with different attitudes operating identical machines side by side on the same material can have as much as 10–12% difference in their consumption rates (Caterpillar, 2011).

The consumption of diesel engines is 0.12–0.16 litres per horsepower per hour, when no other data is available; these values can be doubled in the case of a petrol engine. Electrical motors have an approximate consumption of 0.80 per kW per hour. In addition, the costs of bringing the fuel to the machine, as well as transport and any corresponding losses, have to be taken into consideration too (Rojo, 2010).

Secondary consumption includes crankcase oil, grease, oil for air filters and oil for transmissions and hydraulic mechanisms. Its consumption depends on the condition of the machine and its engine, the working environment and the quality of the lubricants used. Usually, there is a correspondence between the primary and secondary consumptions, since they are both related to the engine's power rating and the prices of petroleum-associated products. A secondary consumption can be estimated as a percentage of the main consumption: 20% for diesel engines, 10% for petrol engines and 5.5% for electrical motors, when no additional information is provided. Box 6.3 gives fuel consumption values and load factor guides for motor graders and track-type tractors (adapted from Caterpillar, 2011):

Repairs, overhauls and adjustments

Estimating equipment repair costs is the most important factor in the operating costs. It is a highly variable cost and of a great economic importance. It includes all spare parts, materials and labour necessary. These costs do not have a uniform pattern that increases gradually with time and the use of the

Box 6.3   Fuel consumption tables and load factor guides for motor graders and track-type tractors

	Low	Medium	High
Motor graders			
120K	9.0–12.9	12.9–16.7	16.7–20.6
160M	14.6–17.8	17.8–23.1	23.1–33.5
14M	15.7–22.4	22.4–29.1	29.1–39.8
24M	36.0–49.2	49.2–68.1	68.1–83.3
Track-type tractors			
D7E	14.8–20.8	20.8–27.2	27.2–34.5
D8R	22.5–32.0	32.0–41.5	41.5–51.0
D10T	61.0–87.0	87.0–113.0	113.0–139.5
D11T	59.0–84.4	84.4–109.8	109.8–135.1

Note: Consumption in litres per hour

Typical application description relative to work

Low   Light road maintenance. Finish grading. Plant and road mix work. Large amounts of travelling. Pulling scrapers, most agricultural drawbar, stockpile, coal pile applications. No impact. Intermittent full throttle operation.

Medium   Haul road maintenance. Average road maintenance, road mix work, scarifying. Road construction, ditching, loose fill spreading. Land forming, land levelling and elevating grader use. Production dozing in clays, sands, gravels. Push loading scrapers: borrow pit ripping, most land clearing applications. Medium impact conditions. Production land-fill work.

High   Heavy maintenance of hard packed roads with embedded rock. Heavy fill spreading, base material spreading and ditching. Ripping/scarifying of asphalt or concrete. Continuous high load factor. High impact. Heavy rock ripping. Push loading and dozing in hard rock. Working on rock surfaces. Continuous high impact conditions.

machine. They depend on the machine and type of work, the daily mainte-nance (greasing, lubricants, cleaning, etc.) and the correct usage and assembly of the machine, as well as the opportunity for, and quality of, the repairs at the construction site (Nunnally, 2010). Labour costs are a third of the total costs, and the rest is spare parts and materials. A discussion of cost models to fore-cast equipment repair costs can be found in Mitchell et al. (2011).

The cost estimates are usually proportional to the amortisation (Nichols and Day, 1998). Manufacturers also recommend straight-line approaches for establishing repair reserves (Caterpillar, 2011). Non-linear amortisation leads to a higher estimate of the cost of repairs at the start of the working life of the unit. In the case of automotive machines, hourly costs of 60–100% of the real value of the linear hourly amortisation are estimated. In the case of operated or towed units, the value is 40–60%.

The cost of repairs can also be estimated as a percentage of the purchase price (Peurifoy et al., 2010). In the case of intensive tasks it is 40–80%, in normal tasks 30–65% and in easy tasks 25–35%. Nevertheless, these figures are merely for illustration purposes, since some machines used in very intensive tasks can reach up to 150%.

Cost of tyres

Tyre costs are an important part of the hourly cost of any wheeled machine. These costs can represent a third of the total cost in heavy machinery. In some cases, tyres are sold separately, being adjusted to the type of work carried out by the unit. On site, tyres experience a lot of wear, so they are usually replaced every 2500–4000 working hours (30 000 to 50 000 km). In the case of scrapers or loading paddles under very harsh working conditions, the life is reduced to 1000 hours. Since the working life of tyres is less than that of machines, the costs of property and operation are analysed separately. Therefore, the amortisation of these types of units is obtained by deducting the cost of tyres from the cost of acquisition. Its hourly cost is calculated as the relationship between the cost of chambers and tyres and the working life in hours, considering 10% over the cost of repairs (i.e. flat tyres). Tyres show evidence of predictable patterns that are worthy of investigation in their own right (Vorster, 2009).

Operators' wages

This type of cost is related to the costs of personnel required to operate the machine. It includes the operator and in some cases their assistants. This cost depends on the local labour conditions. The complexity of some units and the impact of the machinery on production costs requires specialists who have to be paid higher salaries than those in labour agreements. The operator's wages must be in accordance with the rest of the company, labour market, skills, characteristics and category of the position, as well as the actual remuneration. The remuneration of the personnel is in two parts: the basic work payment corresponding to the operator's hours of work, whether they are using the machine or not, and an additional payment related to the machine's performance. This cost is distributed among the work hours of the machine, which are normally two thousand hours per year.

The unit cost of labour for the machine's owner is based on the following (Chen and Keys, 2009):

- part established by a labour agreement: includes the basic salary, fixed for each category, assistance and activity bonuses, distance bonuses, holidays and extraordinary payments

- other compensation: extraordinary hours, bonus for dangerous situations, social bonuses and other types of bonuses for productivity or working hours of the machine
- other burdens: includes travelling or training expenses
- social security etc.: social security, unemployment, professional training, salary guarantee fund, accidents and so on. These are approximately 30–40% of the operator's salary.

6.5 Maintenance

Maintenance comprises a set of activities aimed at ensuring that the machine is fully available for its operation at minimum cost, with maximum performance, and under optimum operation and safety conditions. Correct maintenance entails a high degree of reliability, thus the equipment will have minimum costs of production associated to it. These operations are carried out on both active and inactive units, in order to avoid or solve failures and breakdowns and to ensure that the equipment is available whenever required. Maintenance operations must be quick, efficient and cheap. The main objectives of maintenance tasks are the following (Bloch and Geitner, 2004):

- reduce the costs related to stops caused by accidental failures or breakdowns of the machinery to the minimum, in the case these lead to production or service losses, including the costs corresponding to the maintenance itself
- limiting the deterioration of the machine and, as a consequence, the increase in failures or decrease in the product's quality.

Each service maintenance event contains a group of tasks based on specified maintenance time intervals. The maintenance planning tasks will depend on the size of the company, complexity of equipment, number of identical machines, nature of the operations, cost of stops and so on. A procedure is required to avoid or at least reduce the number of failures or breakdowns and to detect and diagnose the defects or to repair and correct usage problems, taking into account the budget.

There is no rigid classification for maintenance systems, so that each company can choose the most appropriate system for each type of machine. Some machines require advanced preventive maintenance systems while other units operate until a failure is detected. Generally speaking, maintenance systems can be broken down into preventive-overhaul maintenance and failure-corrective maintenance. The first takes into account scheduled maintenance in order to prevent a component from failing during the equipment operation time (e.g. systematic inspection, detection, component rebuilding or replacement based on a specified time interval). The second one can be further divided into two more specific categories: rebuild maintenance, which retains a component at a specified level of performance by rebuilding it based after a certain time period; and replacement maintenance, which replaces a specified component with a new one to preserve the fulfilment of the machine's overall performance. Generally, a component has to be replaced with a new one after a certain number of rebuilds. If possible, the failure-corrective maintenance should be completely eliminated by a properly scheduled preventive-overhaul dictated by the machine's cost and downtime (Chen and Keys, 2009).

Maintenance policies can be classified as follows (Bloch and Geitner, 2004):

- repair of failures: machines are in operation until a fault is detected. This is repaired as soon as possible. This does not involve the omission of first-level maintenance. This system is used in small companies where specific staff for these tasks cannot be justified, outsourcing the repairs. Despite the apparent economy of this type of service, it is only justified on a few occasions or when there are many machines with a vast capacity. The economic problem caused by the sudden and unexpected stop of a machine can lead to stopping other equipment that depends on these units
- routine maintenance: general instructions are provided for the maintenance of homogeneous groups of machines to avoid faults. The frequency of these tasks is usually based on the common-sense and experience of the maintenance supervisor. These tasks include greasing, tests, inspections and adjustments. It is a low-cost system, given its simplicity, which can solve many faults before they occur
- planned preventive maintenance: established in revision and replacement cycles for the most important elements of the machine, relying on the manufacturer's instructions, following the rate of use, construction site and so on. This method enables the detection of faults, frequencies, damaged parts and so on, which lengthens the working life of the elements. At the planned time, the part or set is replaced, even if it is still in good working order.

Preventive maintenance is more expensive in the short term, but it can be used to schedule the times out of service and avoids catastrophic failures, thus increasing the efficiency of the service. The repair of failures is more expensive in the medium and short term, since it does not enable the scheduling of stoppage times, increasing the probability of detecting major faults and decreasing the efficacy of the repair service. Preventive maintenance is used to analyse the machinery's wear, and repair it before important and costly failures and breakdowns occur; therefore, preventive maintenance presents important advantages (Barlow and Hunter, 1960):

- it is faster: minor failures, saving repair time, possibility of programming the repairs, and smaller impact on the work situation of the machine
- it is cheaper: shorter repair time, and savings in labour and the number of parts replaced.

Nevertheless, planned preventive maintenance activities do not detect failures with the maximum anticipation possible and do not solve the causes that lead to failures. Maintenance should be understood from the design or acquisition point of view. Manufacturers are part of the design of their products and can take the necessary measures to ensure that they are robust and, thus, decrease the number of maintenance tasks and costs to the bare minimum.

The final purpose and specifications of maintenance management activities is described in so-called total productive maintenance (TPM) (Nakajima, 1988). It is designed to maximise the efficiency of the equipment (improving its global efficiency), establishing a productive maintenance system with a broad scope that covers the whole working life of the equipment with the participation of all employees, starting with the top level management down to the operators. Its purpose is to promote productive maintenance with the 'management of motivation' or the activities of small groups of volunteers. It is a philosophy that prioritises management activities focused on teams, and is based on the efficient training of employees, ensuring that they are trained to understand the basic information and key components of the equipment they are using. It adopts a medium- and long-term approach in order to achieve the desired results, working in coordination with other departments of the company and management areas. The twelve steps of TPM, as stated by Nakajima (1988) are explained in Box 6.4.

Box 6.4   Implementation of total productivity maintenance (TPM)

1. Announce top management's decision to introduce TPM.
2. Launch educational campaign.
3. Create organisations to promote TPM.
4. Establish basic TPM policies and goals.
5. Formulate a master plan for TPM development.
6. Hold TPM 'kick-off'.
7. Improve equipment effectiveness.
8. Establish an autonomous maintenance programme for operators.
9. Set-up a scheduled maintenance programme for the maintenance department.
10. Conduct training to improve operation and maintenance skills.
11. Develop initial equipment management programme.
12. Implement TPM fully and aim for higher goals.

References

Barlow, R. and Hunter, L. (1960). Optimum preventive maintenance policies. *Operations Research*, 8(1), 90–100.

Bloch, H.P. and Geitner, F.K. (2004). Machinery component maintenance and repair. In: *Practical Machinery Management for Process Plants*, Vol. 3 (3rd Edition). Elsevier, Amsterdam.

Caterpillar (2011). *Caterpillar Performance Handbook* (41st Edition). Caterpillar Inc., Peoria (Illinois).

Chen, S. and Keys, L.K. (2009). A cost analysis model for heavy equipment. *Computers & Industrial Engineering*, 56(4), 1276–1288.

Cross, T.L. and Perry, G.M. (1995). Depreciation patterns for agricultural machinery. *American Journal of Agricultural Economics*, 77(1), 194–204.

Day, D.A. and Benjamin, N.B.H. (1991). *Construction Equipment Guide* (2nd Edition). Wiley, New York.

Douglas, J. (1975). *Construction Equipment Policy*. McGraw-Hill, New York.

Fan, H., AbouRizk, S., Kim, H. and Zaïane, O. (2008). Assessing residual value of heavy construction equipment using predictive data mining model. *Journal of Computing in Civil Engineering*, 22(3), 181–191.

Gransberg, D.D., Popescu, C.M. and Ryan, R. (2006). *Construction Equipment Management for Engineers, Estimators, and Owners*. CRC Press, Boca Raton (Florida).

Halpin, D.W. and Woodhead, R.W. (1998). *Construction Management* (2nd Edition). Wiley, New York.

Lucko, G., Anderson-Cook, C.M. and Vorster, M.C. (2006). Statistical considerations for predicting residual value of heavy equipment. *Journal of Construction Engineering and Management*, 132(7), 723–732.

Mitchell, Z., Hildreth, J. and Vorster, M.C. (2011). Using the cumulative cost model to forecast equipment repair costs: two different methodologies. *Journal of Construction Engineering and Management*, 137(10), 817–822.

Nakajima, S. (1988). *Introduction to TPM: Total Productive Maintenance*. Productivity Press, Cambridge, Massachusetts.

Nichols, H.L. and Day, D.A. (1998). *Moving the Earth. The Workbook of Excavation* (4th Edition). McGraw Hill, New York.

Nunnally, S.W. (2010). *Construction Methods and Management* (8th Edition). Prentice Hall, Upper Saddle River (New Jersey).

Peurifoy, R.L., Schexnayder, C.J., Shapira, A. and Schmitt, R. (2010). *Construction Planning, Equipment, and Methods* (8th Edition). McGraw-Hill, New York.

Rojo, J. (2010). *Manual de Movimiento de Tierras a Cielo Abierto*. Fueyo Ed., Madrid (in Spanish).

Stewart, L. (2006). Giants 2006 Listings. *Construction Equipment*, 109(9), 47.

Vorster, M.C. (2009). *Construction Equipment Economics*. Pen Publications, Christiansburg (Virginia).

Further reading

Gransberg, D.D., Popescu, C.M. and Ryan, R. (2006). *Construction Equipment Management for Engineers, Estimators, and Owners*. CRC Press, Boca Raton (Florida).

Peurifoy, R.L., Schexnayder, C.J., Shapira, A. and Schmitt, R. (2010). *Construction Planning, Equipment, and Methods* (8th Edition). McGraw-Hill, New York.

7 Productivity and Performance

7.1 Educational outcomes

The improvement in productivity and performance of construction processes involves a reduction in cost and time, thus enhancing the construction company's competitiveness in the long term. By the end of this chapter, readers will be able to:

- understand the importance of productivity and performance in construction companies
- realise the importance of the work study as a tool for improving the productivity through the analysis of work measurement methods and techniques
- assess the performance of equipment and other factors that affect production
- recognise value engineering and benchmarking in construction companies as tools that can be used to reduce costs and improve the value offered to clients.

7.2 Productivity and performance

Productivity can be defined as the amount of goods and services generated by a productive unit in a unit of time. In general terms, productivity is defined simply as output divided by input (Drewin, 1982). It can be considered as an indicator of how well the units of production (land, capital, labour and energy) are utilised. This concept is crucial for the development of any business activity, since those that do not improve their productivity, compared to that of their competitors, are doomed to disappear. In addition, productivity is decisive for the economic progress of one country and a key concept for guiding the management of any production system, including construction

Construction Management, First Edition. Eugenio Pellicer, Víctor Yepes, José C. Teixeira, Helder P. Moura and Joaquín Catalá.
© 2014 John Wiley & Sons, Ltd. Published 2014 by John Wiley & Sons, Ltd.

systems. Productivity is critically important in the context of construction contracts, both large and small. However, productivity is extremely difficult to measure due to the heterogeneity of construction products as well as of their inputs where many products come from outside the industry (Loch and Moavenzadeh, 1979). A review of the literature on productivity measurement in construction can be found in Motwani et al. (1995).

Techniques of study and measurement of work have demonstrated their efficiency in the industry for improving their productivity. In this context, high productivity means doing the work in a shortest possible time with the least expenditure on inputs without compromising quality and with minimum wastage of resources (Kumar and Suresh, 2008). The construction industry is characterised by its transhumance, limited works units, with a low degree of specialisation, a high number of temporary staff hired, existence of subcontractors and so on. However, this is not an obstacle for the improvement in productivity and the reduction of costs.

An increase in production does not necessarily translate into an increase in productivity. To increase productivity, all processes that constitute a company's activity must be analysed, and their efficiency must be optimised. In accordance with the International Labour Organisation, the direct resources to increase productivity can be summarised as follows (International Labour Office, 2008):

- capital investment:
  - design of new basic procedures or improvement of existing ones
  - installation of more modern machinery or equipment, with a greater capacity or modernisation of existing equipment.
- improved management, through reduction in:
  - work of the product
  - work of the process
  - unproductive time, attributable to management or workers.

Factors influencing productivity are shown in Figure 7.1. They can be classified broadly into controllable and uncontrollable factors (Kumar and Suresh, 2008).

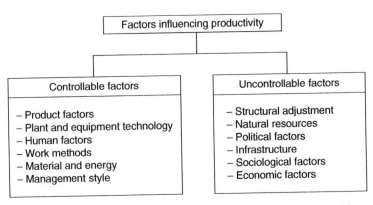

Figure 7.1   Factors influencing productivity (developed from Kumar and Suresh, 2008).

Productivity must not be confused with performance, which is the relation between forecast and executed work, either as a result of the relation with production or with the time needed to carry out an activity. Performance contributes to increasing or decreasing productivity, without modifying the production resources; nevertheless, it influences its efficiency. The loss in productivity (in terms of time of execution) is due to inefficiencies detected in the total time invested throughout the operation due to different causes. Therefore, the time used for activities is broken down in the following (based on Groover, 2007):

- base work content: the quantity of work required to manufacture the product or develop the activity if the project were to be perfect, if the procedure or manufacturing or execution method were fine-tuned or if there were no losses in time attributable to any cause (apart of the pauses given to the producer for rest times). Therefore, the base content of the work is the minimum time of execution
- unnecessary work: additional work due to bad design or poor specifications of the product, or inefficient production or operating methods
- ineffective or unproductive time, due to deficiencies in management or attributable to the worker. The ability and motivation of workers as well as the competence and attitudes of managers have a significant bearing on productivity.

Box 7.1 (adapted from AACE, 2004) reflects several issues regarding measurement and allocation of responsibility when a loss in productivity occurs.

Box 7.1   Measurement and allocation of responsibility

- field crews are not completing work tasks unless a contractor has a good productivity monitoring plan
- there is no way to measure productivity in a contemporaneous manner. Thus, productivity losses can be difficult to prove with the degree of certainty demanded by many owners
- lost productivity is often evaluated at the end of a project. Consequently, often only a gross approximation or a total cost estimate can be made
- there is no common way to calculate lost productivity among professionals
- low confidence in the resulting analysis can be due to fact that the quality of the results of some methods is not always repeatable
- once lost productivity is calculated, it is difficult to establish causation. This situation is frequently a matter for dispute between owners, contractors and subcontractors.

7.3 Work study

Work study refers to the group the techniques used to examine human tasks in any context, and which systematically lead to the research of the factors that have an impact on the efficiency and economy of the situation analysed, with the aim of implementing improvements (BSI, 1992). Work study is focused specifically to suit construction applications (Harris et al., 2006), and it aims at improving the existing and proposed ways of doing work and establishing standard times for work performance (Kumar and Suresh, 2008). It also attempts to lead to higher productivity with no, or little, extra capital investment, and involves two distinct yet interrelated techniques. The first, method study, is concerned with establishing the most effective manner of performing a specific job; it seeks to improve production. The second, work measurement, involves finding how much time is required to execute the work, in order to reduce of any inefficient time associated with the job; it aims to estimate the efficiency. There is a close relationship between work study, method study and work measurement as Figure 7.2 shows.

Some of the benefits of using work study methods are (Kumar and Suresh, 2008):

- increased productivity and operational efficiency
- reduced cost of the product by eliminating waste and unnecessary operations
- improved workplace layout
- reduced material handling costs and time
- enhanced manpower planning and capacity planning
- establishing fair wages for employees and the basis for a sound incentive scheme
- getting better working conditions and job satisfaction for employees
- improving production flow with minimum interruptions

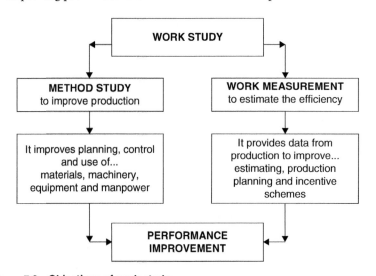

Figure 7.2 Objectives of work study.

- providing a standard of performance or quality that is expected in the production process
- enhancing industrial relations and employee morale.

However, there are problems in applying work study methods in construction (Harris et al., 2006): the temporary nature of most construction projects and some difficulties such as weather, bad ground, dispersed locations, unique tasks in each project, subcontracts works on site and so on. Nonetheless, these methods provide a disciplined procedure to improve productivity and competitiveness; thus they are increasingly finding support in the construction industry.

7.4 Method study

Method study involves the systematic recording and critical examination of the factors and resources involved in existing systems and those proposed for execution, as a means to developing and applying the most effective procedures and reducing costs (BSI, 1992). In short, method study is essentially concerned with finding better ways of doing things. It is also called methods engineering or work design. Table 7.1 shows some of the possible symptoms required for the study of works methods.

The aims of a study of methods are the following (Kumar and Suresh, 2008):

- improvement of processes and procedures
- improvement in the layout at the work site, as well as the design of equipment and facilities
- economising human effort and reducing unnecessary strain
- improving the use of materials, equipment and labour force
- creating improved working conditions.

Method study requires a critical attitude and systematic action to analyse and improve the execution of specific activities. Addressing problems with an open spirit, eliminating preconceived ideas and prejudice, accepting facts only and not opinions, acting on causes and not on effects and considering that a better method is always possible are the general principles that

Table 7.1 Symptoms that show the need for applying the method study.

Excess working hours	Bottlenecks in the flow of materials	Excessive waste of materials
Frequent faults in the machinery	Works that lead to physical strain	Badly programmed temporary works
Bad quality in the execution	Delays caused by subcontractors	Subcontractors affected by delays
Shortage of resources	Not enough information	Congested works
Bad working conditions	Excessive costs	High personnel rotation
Delayed programme	Bad distribution at the works	Excessive faults and errors

must govern the study of methods. In order to address and put into practice any type of study aimed at improving methods, five phases are followed:

1. convenient choice of the problem and its definition
2. observing and intake of records of the current method
3. analysis of the current method
4. developing the improved method
5. application and maintenance of the new method.

The adequate choice of work analysed usually leads to selecting the one that provides the greatest profitability, in terms of maximising the benefits of improvement in relation to resources needed. Therefore, the selection usually includes: bottlenecks, unnecessary movements that do not add value to the product, work that requires a high quantity of labour force or machinery or operations that require a repetitive type of work.

The way of critiquing current activities involves the systematic approach to issues on each of the factors included in the method observed and analysed. The technique of answering questions would lead to posing the following (Calvert et al., 1995):

- what activity is carried out exactly? And why is it carried out?
- where is it carried out? And why is it carried out there?
- how much is carried out? And why is such a quantity carried out?
- who carries it out? And why must it be executed?
- how is it carried out? And why is it carried out?
- when is it carried out? And why is it carried out at that moment?

A new method of work can be designed, using these following steps (Harris et al., 2006): eliminate unnecessary work, combine operations or phases of operation, change the order of execution of operations or simplify the operations required. Different tools can be used, such as flow process chart, string diagram, operation process chart, planimetric flow chart or multiple activity chart.

Before adopting a new method, management must approve it, requiring the drafting of a report that includes the costs and general expenses of both methods, as well as the estimated cost-cutting, costs of implementation of the new method and executive decisions required to apply the new method. Finally, after implementing the new method, periodic verifications are required, more frequently at the start, to ensure that the activity is being carried out accordingly. These controls are spaced in time until the normal supervision system is achieved.

7.5   Work measurement

There is almost unquestioning dependence on the historical value of time in construction. In fact, time is related to the measurement of productivity. Work measurement is defined as the application of the techniques designed to establish the time that a qualified worker takes to complete a specific task, in accordance with a pre-established method (BSI, 1992). A qualified worker is one who has acquired the skills, has the know-how and is physically able to carry out the work safely and in compliance with the applicable quantitative and qualitative requirements. Work measurement does not pretend to engineer the job to be optimal; it only evaluates the time associated with a particular construction method. Also, work measurement enables productivity comparisons between similar activities, distinct construction crews, alternative plant or equipment, before and after relevant events, and so on.

The first objectives of work measurement are to determine unproductive periods and their causes, and to find ways of eliminating them. Accordingly, methods required to eliminate idle time or to optimize working time are used as auxiliary tools for work measurement. The following phases are established to measure the work performed (adapted from Groover, 2007):

- break down the working time of all elements
- measure the working time of the elements while estimating the speed and precision
- calculate the normal time for each element or level
- estimate the increasing coefficient for each element
- obtain the cycle for each resource
- calculate the saturation of each resource on the equipment.

Activity sampling is one of the best ways of obtaining a detailed knowledge of any production process performance. This is based on statistical techniques that allow overall activity rates to be inferred from a limited number of observations (Winch and Carr, 2001). Two of the techniques used for work measurement are:

- time study: appropriate for highly systematised and repetitive works, carried out by one or a few resource units
- work sampling: covers the rest of the possible scenarios, such as works that are not systematised, that have long cycles or that are carried out with a wide range of resources.

Time study is a sampling technique that consists of separating a job into measurable parts, with each element timed individually. Time study precision depends on the size of the sample: in general, a minimum of 30 measurements are taken, with more observations also carried out to include working at different hours of the day and on different days. Some studies entail continuous timing with an electronic stopwatch or video camera.

Work sampling is based on checking whether a resource is stopped or is working at random times. Statements about how time is spent during the activity are made from these observations. The working and stoppage time can be estimated, as well as its statistical error, based on a binomial probability distribution. A run can be carried out when a set of resources is observed and the working or stoppage times for each unit is noted down. To plan the observations correctly, all activities must be observed a number of times, proportional to their duration. In addition, the observations should be taken over reasonably long periods (e.g. two weeks or more), but rigid rules will need to bend with the situation and the design of the study (Ostwald, 2001). Box 7.2 shows the calculation of the number of observations required for a given absolute error.

The basic operations or elements of work must be known to time these tasks. They are defined as any part of time with a resource limited by the

---

Box 7.2   Number of observations required for a given absolute error at various value of p, with 95.45% confidence level

The number of observations N required in a work-sampling study is dependent on the activity and desired degree of accuracy. This number can be obtained from the formula $E = \sqrt{\dfrac{p(1-p)}{N}}$ and the required sample $N = \dfrac{Z^2 p(1-p)}{E^2}$ where

E = absolute error
p = percentage occurrence of activity measured
N = number of random observations
Z = number of standard deviations to give desired confidence level. In this table $Z = 2$, for 95.45% confidence.

The next table gives the number of observations needed for a 95.45% confidence level in terms of absolute error.

Percentage of total time occupied by activity, p	Absolute error		
	±1%	±2%	±3%
1	396	99	44
2	784	196	87
5	1900	475	155
10	3600	900	400
20	6400	1600	711
30	8400	2100	933
40	9600	2400	1067
50	10000	2500	1111

For example, if a carpenter is idle for 10% of the time and the designer of the study is satisfied with a 3% range (meaning that the true percentage lies between 7.0% and 13%), the number of observations required for the work sampling is 400. Then, the number of observations to be taken is usually divided over the study period. Thus, if 400 observations are to be made over a 20-day period, observations are usually scheduled at 20 per day.

instants at the start and the end of the resource. The element is the smallest part of an operation observed by work measurement. Table 7.2 shows the classification of these elements.

Most of the above factors are due to the procedures used and the precision and speed of operations performed. This means that the same work, carried out by different operators, will have different performances (even when carried out by the same operator). To make the operators' time usable for all workers, a measure of operation speed or performance rating must be included to 'normalise' the job. The application of a rating factor (see Table 7.3) yields the "normal operation time" (BSI, 1992). Therefore, if this work method is followed step by step, the activity will increase its speed and the precision of its operations.

The rating factor is a judgment made by the observer about the worker's speed and skill. It is a means of comparing the performance of the worker by using experience or other benchmarks. Additionally, a numerical factor is noted for the elements or cycle: 100% is normal; rating factors less than or greater than normal indicate slower or faster performance, respectively (Ostwald, 2001). The optimum performance is defined as that obtained naturally without forcing qualified workers to work above the normal working hours, with the average established in a shift or normal working day, provided that the method specified is respected and that they are motivated to work. Normal activity is defined as that corresponding to three-quarters of the optimum activity. Table 7.3 shows a simple scale representing standard rating.

Approximately 50% of individuals reach the optimum activity when they have incentives, and 98% reach the normal activity levels. Only 2% of operators are capable of reaching an activity that is 1.25 times the optimum. Any increase in activity corresponds to the decrease in time, with a resulting constant product between both. Therefore, a skilled employee responsible for timing tasks must determined the rhythm or activity of operations before the time measurements are taken. Levelling the times measured in an operation involves reducing the times used in normal activity.

Normal time is the time that a normal operator would be expected to take to complete a job without the consideration of allowances. It is found by multiplying a selected time for the element by the rating factor (Niebel, 1993):

$$T_n = T_o \cdot RF \qquad (7.1)$$

where

$T_n =$ normal time, hours, days, etc.
$T_o =$ observed time, hours, days, etc.
$RF =$ rating factor, arbitrary set, number

However, the work developed throughout a normal working day cannot be completed within a normal shift, since some interruptions are required to recover from fatigue, go to the bathroom and other unavoidable delays. Allowances for such pauses are divided into three components (Niebel, 1993): personal, fatigue and delay (PF&D). The fair day work standard or standard

Table 7.2   Classification of elements or elemental operations.

Cycle	Regular or repetitive	Only once in every work cycle
	Irregular or frequency	Once every number of cycles
	Casual or strange	Not part of the work
	Interior	Carried out by the operator when the machine is in automatic work mode – these elements do not modify the duration of the work cycle
	Exterior	Executed by the operator when the machine is stopped – these elements are part of the work cycle
Executing party	Manual	Executed by the operator during the work cycle
	Mechanical	Used by the machine to execute an element in the automatic operating mode
Duration	Constant	With an invariable value (or practically the same)
	Variable	With a duration that depends on the value of a determined parameter

Table 7.3   Rating scale for varying levels of worker performance (developed from BSI, 1992).

Scale	Description of the activity
125	Very fast, high skill and motivation required
100	Active, specialised skill, motivation
75	Not very fast, average skill, not too much interest
50	Very slow, no motivation or skill

time is calculated by taking the normal time and adding the PF&D allowances. In the standard time, a skilled worker with a normal activity would achieve a standard performance throughout a shift or day of work. PF&D allowances are generally in the range of 10–20%. An allowance multiplier is found using:

$$F_a = \frac{100\%}{100\% - PF \& D\%} \tag{7.2}$$

where

F_a = allowance multiplier for PF&D, number
$PF\&D$ = personal, fatigue, and delay allowance, percentage

Finally, the time required by a trained and motivated worker to perform construction tasks while working at a normal tempo, standard time H_s, can be calculated by using:

$$H_s = T_n \cdot F_a \tag{7.3}$$

Figure 7.3 summarises the whole process. Box 7.3 gives a guide to the number of cycles to be observed in a time study (adapted from Niebel, 1993). Given the variability of construction activities, the difference between the standard

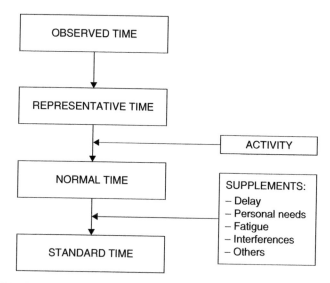

Figure 7.3   Determination of the standard time of a task.

Box 7.3 Guide to the number of cycles to be observed in a time study			
When time per cycle is more than	**Minimum number of cycles of study (activity)**		
	Over 10 000 per year	**1000–10 000**	**Under 1000**
8 hours	2	1	1
3 hours	3	2	1
2 hours	4	2	1
1 hour	5	3	2
30 minutes	8	4	3
5 minutes	20	10	8
1 minutes	40	20	15
0.1 minutes	120	60	50

and normal time can be great, so that the databases usually state normal times, to which the user must apply the adequate unforeseen circumstances to each case. A realistic planning period is usually double the normal time.

7.6   Equipment performance

Measuring equipment performance is critical to any construction business success. However, the production of a machine depends on many different factors, such as the weather, complexity of the work, shifts, state of the machinery, dimensioning of units, skill and experience of the operator, and the existence of production incentives (Alfeld, 1988). There are manuals, technical files, tables and graphs available with data on the theoretical production of each unit. Nevertheless, these figures have been calculated or measured under special and specific circumstances, that is, usually under favourable or optimum conditions.

The standard production rate P of a piece of construction equipment is defined as the number of output units per unit time. Therefore, the production can be defined by real working hour, by hour available, by hour of use or by hour of productive work. This production can be considered daily, monthly or total.

While the standard production rate of a piece of equipment is based on 'standard' or ideal conditions, equipment productivities at job sites are influenced by actual work conditions and a variety of inefficiencies and work interruptions. To correct the theoretical production P_T and forecast the real production of a unit P_R, the first must be multiplied by a series of factors of production. Each of these various adjustment factors must be determined from experience or observation of job sites (Hedrickson and Au, 1989). However, most manuals and technical files provided by equipment manufacturers add tables or data that can be used to calculate the specific production factors as follows:

$$P_R = P_T \cdot f_1 \cdot f_2 \cdot \ldots \cdot f_n \tag{7.4}$$

Also, equipment can only be operated during the working calendar, with a real working time of H_p, due to unforeseen circumstances, such as weather conditions, strikes, natural hazards and other unwanted events. The machine is in operation, in good working order and prepared for work during the time of operation H_d. When the machine is out of working hours H_m, estimated tasks are used, such as maintenance and other hours H_a, which cannot be forecast, such as fault repairs.

The equipment in operation can be stopped H_p hours for unforeseen reasons, due to the deficient organisation of the works, the lack of work planning, incorrect supply planning, incorrect dimensioning of equipment, faults in other machines and so on. Therefore, a machine only has a valid working time of H_u, which can be used to produce during H_t hours, or carry out other non-productive or complementary activities, such as changes or the preparation of sites during H_c hours.

A construction site manager must know the factors that have an impact on the performance of the machinery. This can lead to action to solve faults or raise the productivity values. These factors are (Rojo, 2010):

- availability factor F_d: relation between the time available and real working time. If the value is low, the causes must be studied: poor maintenance, slow repairs, lack of spare parts, machine in bad condition or not enough reliability:

$$F_d = \frac{H_d}{H_l} \tag{7.5}$$

- usage factor F_u: relation between the available and used times. It indicates the quality of organisation and planning of the works. A low value can be due to bad scheduling of tasks, lack of communication between managers and supervisors, unforeseen storage space needs and so on.:

$$F_u = \frac{H_u}{H_d} \qquad (7.6)$$

Another two complementary indices can be used. The stoppage index p defines the relation between interruptions due to the organisation of the works, bad machine coordination, stoppage due to faults of other machines and so on, and real working time:

$$p = \frac{H_p}{H_l} \qquad (7.7)$$

The usage factor F_a defines the quotient between the usage time of a machine and the real working time:

$$F_a = \frac{H_u}{H_l} \qquad (7.8)$$

As an example, Box 7.4 shows a procedure for pile installation control, considering productivity factors applied to the foundation of highway bridges (adapted from Zayed and Halpin, 2005, and Peurifoy et al., 2010).

Box 7.4   Procedure for pile installation control considering productivity factors

Bored piles are widely used in the foundation of highway bridges. The steps for executing a concrete bored pile can be summarised as follows:

- adjust the piling machine on the pile axis
- lower the auger to the drilling depth
- start drilling until the auger is filled
- return from the drilling depth up to the top of the pile hole
- swing to the unloading area
- unload the dirt in the unloading area
- swing back to the top of the hole
- repeat steps 2–7 until the pile is completely drilled
- relocate the machine and start steps 1–8
- start erecting the rebar cage using a crane
- erect the concrete pouring tool, either funnel or tremie, into the hole
- use funnel for dry method and tremie for wet method
- start pouring the concrete and finishing the pile.

(Continued)

However, a large number of factors affect the productivity and cost estimation processes:

- insufficient statistical samples around the foundation area
- soil types (sand, clay, etc.) differing from site to site due to cohesion or stiffness
- natural obstacles and subsurface infrastructure construction obstacles
- drill type (e.g. auger bucket)
- method of spoil removal, size of hauling units and site restrictions
- pile axis adjustment
- equipment operator efficiency
- concrete pouring method and efficiency
- lack of experience of rebar crew and method of pouring
- waiting time for other operations (e.g. pile axis adjustment)
- weather conditions
- cycle time
- job and management conditions.

7.7 Assessment of production/productivity

The success of a construction project, from a financial point of view, is highly dependent on the achievement of a certain level of labour productivity in performing the job. However, many events and circumstances, such as changes in the project, excessive overtime, resources shortages, rework or errors and severe weather negatively affect productivity. These reasons justify measuring the productivity on construction sites. However, this is a complex task because it is necessary to find measures of productivity that could be used to evaluate site performance (Ramirez et al., 2004). The construction industry is a project-oriented industry; thus, it is difficult to obtain a standard method of measuring construction labour productivity, owing to the uniqueness and non-repetitive operations of construction projects. In fact, job-site productivity can only be estimated under a specific set of work conditions either for each craft or each type of construction. Thus, a set of work conditions specified by the owner or contractor defines the base labour productivity. However, we can define the labour productivity index as the ratio of the job-site labour productivity under a different set of work conditions to the base labour productivity.

Construction labour productivity is typically measured as output per hour. The daily standard production rate or standard performance could be defined as the production throughout a work shift or day by a qualified worker with a normal activity (Rojo, 2010). In the case of machinery, standard hourly production can be defined as that obtained during 54 uninterrupted minutes with that production, following a particular work method on a site with

Table 7.4   Standard production forecasting methods (developed from Rojo, 2010).

Method	Characteristics	Applications
Comparison with similar works	Low precision. Dangerous even for experienced technicians.	Selection of work methods and equipment. Simple works units with a short time of execution.
Tables and graphs	Acceptable precision. Use of the adequate correction factors.	Continuous production equipment. Repetitive and short-cycle units that do not require major transfers during the works.
Detailed calculation of the works cycle	Acceptable precision. More laborious studies. Computer equipment can be used.	Transport equipment. Short cycle units with transfers. Machines or units in series with execution phases. Units used in repetitive works.
On-site testing	Maximum precision. In general, only applicable to works in progress.	All types of equipment and machines. Very common for the comparison of methods, classification units, etc.

particular circumstances and with an average operator skill. Hours with 54 minutes are used, since 6 minutes are estimated as unaccounted work losses. This reduced hour is the estimated duration of the average valid work of a unit under the given conditions. Therefore, any graph or table that offers a production for a machine will be defined as standard production.

This definition determines the specific conditions where the production takes places with precision. There are no formulae to enable the precise forecasting of production with different conditions. The problem is solved with approximations, so that once the standard production and its conditions are known, the adequate production factors are determined to estimate real production. Therefore, standard production is simply a theoretical type of production for determined conditions.

Production per hour of productive work at a specific construction site can be related to standard production P_t with hourly efficiency, the efficacy factor or an operational factor F_e. It is defined as the quotient between the average production per hour of use and standard production of the machine (Rojo, 2010). The non-productive hours of work dedicated to other tasks are considered, such as transfers, preparation of the construction site or auxiliary tasks. It mainly depends on the selection of personnel and the method of work. The value usually ranges between 1.2 and 0.8.

There are different methods used to estimate the standard production of a unit. These methods, their characteristics and applications, are explained in Table 7.4.

7.8   Benchmarking and construction productivity improvement

As world competition intensifies, leading construction companies have raised the awareness for performance measurement among the majority of construction organisations. Thus, benchmarking is a suitable tool for guidance

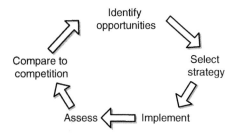

Figure 7.4   Benchmarking as a continuous improvement cycle.

to measure construction performance. Benchmarking can be defined as a technical-competitive analysis tool for products, services or processes. It is an integral part of the continuous improvement cycle (Figure 7.4) as well as being a systematic process used to evaluate products, services and methods, and comparing them with the best practice identified in a group of companies (Ramirez et al., 2004). Measurement, comparison, learning and improvement are the essence of benchmarking. This technique can be very useful to construction companies, whereby the comparison of construction procedures and forms of organisation at the construction site can be highly efficient, since this is an industry where products, processes and equipment are constantly changing.

Therefore, the objective of benchmarking is to optimise results. It mainly involves gathering, adapting and implementing tested methods that have provided positive and revolutionary results in other companies. It is necessary to understand the development of this process and the measures adopted to attain a high degree of performance. It involves having an in-depth understanding of the factors that make this improvement possible. This technique allows savings in costs of approximately 30% in manufacturing and service industries, so that it is reasonable to find savings of this magnitude that can be applied to the construction industry. However, the applicability of manufacturing production to construction has always been a challenge, and benchmarking is not an exception. The construction industry seems to be reluctant to adopt this tool. Park et al. (2005) have claimed that there is no standardised construction productivity measurement; Mohamed (1996) highlighted the lack of relevant conceptual models to support and guide data collection associated with field-based operations, as well as noting the fact that application of benchmarking requires radical changes in the way information is handled and documented.

Benchmarking sets off from the base that it is hard for a company to attain results that are better than those of its competitors in all processes. To analyse these better practices, information is sometimes shared with companies that are not direct competitors, covering similar functions, problems or processes. Therefore, different types of benchmarking can be established, in accordance with different aspects. The most common classification takes into account the relationship existing with a company or organisation participating in the

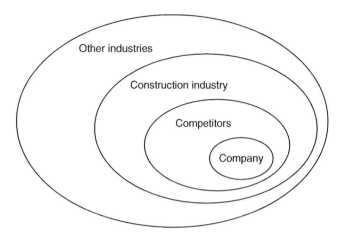

Figure 7.5   Benchmarking out of the box (developed from Andersen, 1995).

study. Therefore, according to the comparison done, benchmarking can be divided into (Andersen, 1995):

- internal: different areas of the same organisation
- competitive: companies within the same sector
- functional: organisations within the same technological area that are not necessarily competitors
- generic: companies from different sectors.

Figure 7.5 illustrates these types of benchmarking. According to Andersen (1995), the further down the list, the further away you go from the company under analysis.

A common error involves considering benchmarking as a mere comparison between data or indicators. It involves identifying, interiorising and adapting the best practices to the company, so that a climate of adaptation to change and constant learning is generated. Most of the studies related to construction productivity and performance measurement are mainly focused on the identification of sources of delays, rather than with analysis of measuring systems and techniques. In the absence of data and measuring methods, construction benchmarking should be focused on three levels: internal, project and external (Mohamed, 1996).

In internal benchmarking, organisations attempt to learn from their own current processes and practices. This method is used when comparing business operations with those of others who do it well, leading to a better understanding of the organisation's processes, resources and functions. To be effective at internal benchmarking, a company needs to record and display the current work processes and practices objectively, identifying weaknesses as well as the best practices within the organisation. It is necessary to establish well-defined indicators and measuring systems in order to identify the 'best-practice' in

Table 7.5   Examples of generic benchmarking in construction companies.

Problem	Compared with
Truck bunching or queuing	Hotel receptions
Planning the delivery of fresh concrete	Hot pizza delivery
Billing process of a construction company	Billing process of a telecommunications company
Production control in precast concrete plants	Production control in automotive manufacturing plants

the organisation (Mohamed, 1996). Then, the non-value-adding activities embedded within the process should be eliminated, in order to get a more efficient practice. After a process comparison against one which does it better, implement changes and monitor the output in order to focus available resources on those aspects that need restructuring. The next step is to benchmark the performance of the construction company against that of another organisation of reference (a better one) in order to increase its productivity and competitiveness.

In project benchmarking, a contractor assesses the performance of construction projects. This type of benchmarking is highly dependent on the project characteristics. Performance measurements have usually focused in terms of complying with completion time and allocated budget (Mohamed, 1996). At this level, for each value-adding activity of the project, relevant customer requirements must be clearly identified and satisfied with the outcome. Also, a set of project benchmarking measures are required to assess their performance, assess their productivity rates and validate their initial estimations.

In external benchmarking, the construction company attempts to identify innovative processes and managerial and technological breakthroughs, developed by other industries. Gable et al. (1993) claimed that practitioners are more receptive to new ideas when these ideas do not originate in their own industry. Table 7.5 shows some examples of generic benchmarking that can accelerate improvement and change through generic benchmarking.

Despite those levels of benchmarking, there are some intrinsic problems to the construction industry, which can hinder the good performance in the adoption of benchmarking within the industry (Ramirez et al., 2004). First, each project is unique, designed for a determined location. Another problem involves the lack of practice in identifying the best practices and, above all, measuring the process indicators. In addition, there are not many examples of benchmarking in the construction industry with results that can be applied to other cases. Nevertheless, the potential improvements in processes, and the reduction in costs and potential deadlines are so high that any effort in implementing a benchmarking process in construction companies is profitable in the long term and significantly improves the quality of decision-making related to the design and construction phases of the facility life-cycle.

References

AACE (2004). *AACE International Recommended Practice No. 25R-03. Estimating lost labour productivity in construction claims*. The Association for the Advancement of Cost Engineering, available 21/12/2012 at: http://www.sde-us.com/docs/emailblasts/JetBlue/LostLabour.pdf.

Alfeld, L.E. (1988). *Construction Productivity: On-Site Measurement and Management*. McGraw Hill, New York.

Andersen, B. (1995). *The Results of Benchmarking and a Benchmarking Process Model*. Norwegian Institute of Technology, University of Trondheim, Trondheim (Norway).

BSI (1992). *BS 3138: Glossary of Terms Used in Management Services*. British Standards Institution, London.

Calvert, R., Bailey, G. and Coles, D. (1995). *Introduction to Building Management* (6th Edition). Routledge, London.

Drewin, F.J. (1982). *Construction Productivity*. Elsevier, New York.

Gable, M., Fairhurst, A. and Dickinson, R. (1993). The use of benchmarking to enhance marketing decision making. *Journal of Consumer Marketing*, 10(1), 52–60.

Groover, M.P. (2007). *Work Systems: The Methods, Measurement & Management of Work*. Prentice Hall, Upper Saddle River (New Jersey).

Harris, F., McCaffer, R. and Edum-Fotwe, F. (2006). *Modern Construction Management* (6th Edition). Blackwell, Oxford.

Hedrickson, C. and Au, T. (1989). *Project Management for Construction. Fundamental Concepts for Owners, Engineers, Architects and Builders*. Prentice Hall, Upper Saddle River (New Jersey).

International Labour Office (2008). *Skills for Improved Productivity, Employment Growth and Development. Report V*. International Labour Conference, Geneva.

Kumar, S.A. and Suresh, N. (2008). *Production and Operations Management* (2nd Edition). New Age International, New Delhi (India).

Loch, J.A. and Moavenzadeh, F. (1979). Productivity and technology in construction. *Journal of the Construction Division ASCE*, 105(4), 351–366.

Mohamed, S. (1996). Benchmarking and improving construction productivity. *Management & Technology*, 3(3), 50–58.

Motwani, J., Kumar, A. and Novakoski, M. (1995). Measuring construction productivity: a practical approach. *Work Study*, 44(8), 18–20.

Niebel, B.W. (1993). *Motion and Time Study* (9th Edition). Irwin, Homewood (Illinois).

Ostwald, P.F. (2001). *Construction Cost Analysis and Estimating*. Prentice Hall, Upper Saddle River (New Jersey).

Park, H.S., Thomas, S.R. and Tucker, R.L. (2005). Benchmarking of construction productivity. *Journal of Construction Engineering and Management*, 131(7), 772–778.

Peurifoy, R.L., Schexnayder, C.J., Shapira, A. and Schmitt, R. (2010). *Construction Planning, Equipment, and Methods* (8th Edition). McGraw-Hill, New York.

Ramirez, R.R., Alarcón, L.F. and Knights, P. (2004). Benchmarking system for evaluating management practices in the construction industry. *Journal of Management in Engineering*, 20(3), 110–117.

Rojo, J. (2010). *Manual de Movimiento de Tierras a Cielo Abierto*. Fueyo ed., Madrid (in Spanish).

Winch, G. and Carr, B. (2001). Benchmarking on-site productivity in France and the UK: a CALIBRE approach. *Construction Management and Economics*, 19(6), 577–590.

Zayed, T. and Halpin, D. (2005). Pile construction productivity assessment. *Journal of Construction Engineering and Management*, 131(6), 705–714.

Further reading

Halpin, D.W. and Woodhead, R.W. (1998). *Construction Management* (2nd Edition). Wiley, New York.

McCabe, S. (2001). *Benchmarking in Construction*. Blackwell, Oxford.

Oglesby, C.H., Parker, H.W. and Howell, G.A. (1989). *Productivity Improvement in Construction*. McGraw-Hill, New York.

8 Quality, Innovation and Knowledge Management

8.1 Educational outcomes

An adequate management of construction processes guarantees compliance with the quality requirements established by the owner. It is also the starting point for a policy of continual improvement and innovation that guarantees the competitiveness of construction companies in the long term. Therefore, by the end of this chapter readers will be able to:

- understand the concepts of quality, innovation and knowledge
- recognise the importance of innovation and knowledge as key elements in the competitiveness of construction companies
- differentiate quality control and quality assurance concepts
- distinguish between certificates and technical approvals and understand their application to construction products and construction systems.

8.2 Quality, innovation and knowledge

The key objective of any business is its survival, and in order to survive, companies have to obtain competitive advantages. Like any other kind of organisation, construction companies must be competitive to survive in an increasingly globalised market. This, however, is not a simple task. The construction industry has peculiarities that clearly distinguish it from manufacturing industry (see Chapter 1). Perhaps the most important of these is that its companies produce and manage by projects. These involve temporary coalitions of different organisations coming together to meet particular targets for a given period. However,

Construction Management, First Edition. Eugenio Pellicer, Víctor Yepes, José C. Teixeira, Helder P. Moura and Joaquín Catalá.
© 2014 John Wiley & Sons, Ltd. Published 2014 by John Wiley & Sons, Ltd.

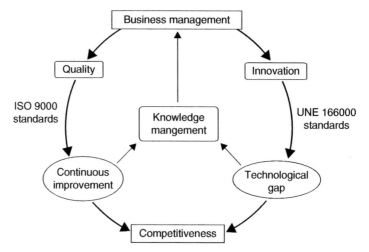

Figure 8.1   The relationship between quality, innovation and knowledge (Pellicer et al., 2008) (Reproduced with permission from Pontificia Universidad Católica de Chile).

there are similarities between the two industries (McCrary et al., 2006): both industries exist to make products and they are part of the (local/regional/global) economy, seeking to survive and striving to make a profit.

Quality, innovation and knowledge are deemed essential for organisations for achieving the ability to adapt quickly to changes within the environment. An effective utilisation of knowledge in an organisation can provide competitive advantages. As shown in Figure 8.1, this important resource is part of a basic feedback process of quality management in the firm, as well as being part of the process of innovation (Pellicer et al., 2008). Thus, it is argued that learning is the highest form of adaptation thereby raising the probability of survival (Oliver, 2008). Nevertheless, although quality and innovation are processes open to standardisation, knowledge management still is not open to standardisation, which prevents improvement in competitiveness, especially in the case of firms that manage and produce by projects.

The construction sector is knowledge-intensive because its activities require a high level of expert knowledge and know-how to solve problems (Dave and Koskela, 2009). A significant part of this knowledge comes from information resulting from continuous improvement and detection of technological gaps; these are a consequence of the quality and innovation systems of the company. However, the temporary nature of the construction activities makes it difficult to transmit knowledge from one project to another owing to the continuous reorganisation of project teams. For this reason, employees and collaborators must be considered the most important part of a firm, and companies ought to provide the best possible environment for knowledge to be appropriately developed. These are factors which make a difference in terms of the competitiveness of firms (Castro et al., 2012).

8.3   Quality control

Quality control is the part of quality management that ensures that product and service requirements are complied with. It facilitates the measurement of the quality characteristics of an activity or a product, compares them with the established standards and analyses the differences between the results obtained and the desired results, in order to make decisions which can correct these differences. It may include whatever actions a construction business deems necessary to provide for the control and verification of the legal, aesthetic and functional requirements of a project.

The verification of quality requirements in the construction process is mainly based on quality control. Technical specifications define the type of controls that must be carried out in each activity to guarantee that the construction project is correctly performed. They include not only materials, but also the execution and completion of the construction project. To this end, Chan and Wu (2002) provide a literature review of quality function deployment (QFD), which is an effectiveness tool that provides a means of translating customer requirements into the appropriate technical requirements. The same philosophy is applicable to the drafting of design projects by consulting firms. A quality control plan can be implemented to ensure that construction procedures are performed in compliance with specifications and requirements under contract. Quality control reduces the possibility of changes, errors and omissions, which in turn results in fewer conflicts and disputes (Juran and De Feo, 2010).

One way of controlling quality is based on the inspection or verification of finished products. The aim is to filter the products before they reach the client, so that products that do not comply with the requirements established are discarded or repaired (Juran and De Feo, 2010). This reception control is usually carried out by people who were not involved in the production activities, which means that costs can be high, and preventative activities and improvement plans may not be taken into account (Al-Tmeemy et al., 2012). It is a final control, located between producer and client, and although it has the advantage of being impartial, it has a large number of drawbacks such as: slow information flows; inspectors not familiarised with the circumstances of production; and responsible parties for quality not involved in production.

However, a complete inspection of all the units produced can be physically impossible when tests are destructive. In these cases, the decision to accept or reject a full batch must be made on the basis of the quality of a random sample. This type of statistical control provides less information and contains sampling risks. Nevertheless, it is more economical, requires fewer inspectors and speeds up decision-making, while the rejection of the whole batch stimulates suppliers to improve their quality.

Statistical control was standardised after World War II, characterised by the consideration of quality characteristics as random variables, for which reason it focuses on manufacturing or production quality. This type of control also identifies the causes of variations and, as a consequence, establishes procedures for the systematic elimination of those causes so that quality can be continually improved (Stuart et al., 1996).

Statistical control can be applied to the final product (acceptance control) or throughout the production process (process control). The statistical controls at reception establish sampling plans with clearly defined acceptance or rejection criteria; complete batches are tested by means of random sampling. The sampling control can be based on the inspection by attributes in line with the ISO 2859 standard, or on the inspection by variables in line with the ISO 3951 standard (Schilling and Neubauer, 2009). As regards the statistical control of processes, different tools can be used to take decisions when the process is out of control, such as control graphs. Likewise, process capacity studies determine whether the processes have the capacity to produce within the limits of the quality specifications contracted.

A construction company should reduce the cost of poor quality as much as possible, and ensure that the result of its processes complies with the requirements agreed with the client (Burati et al., 1992). Therefore, in order to guarantee that the product acceptance control is successful (the so-called external control) the construction company must set up a series of controls in its own production chain to guarantee the quality of its construction activities (defined as internal control).

Both internal and external controls can be carried out by the construction company, by the owner or by an independent organisation. For example, the control of concrete received by the contractor can be carried out by an independent entity; the execution of steelworks can be controlled by the project manager (on behalf of the owner), or the construction company can establish an internal control for the execution of the building work. Despite all this, there is a great potential for quality improvement in the construction process (Arditi and Gunaydin, 1997).

8.4   Quality assurance in accordance with ISO 9001

Quality assurance constitutes a set of planned and systematic actions required to guarantee that the products and services in an organisation comply with specified requirements. It involves more than just checking the final quality of products to avoid defects, as is the case in quality control. Instead, it also includes product quality in a planned way in all the production stages and up to the final delivery to the client. The quality is attested to by the documentation of actions and audits, which demonstrate that all the processes are under control. This involves the development of work and product design procedures to prevent errors from occurring in the first place. This stage involves the development of a quality system based on planning and backed up by quality manuals and tools, which generates quality from the start of the process, instead of waiting until the final stage. However, as Arditi and Gunaydin (1997) claimed, large amounts of time, money and resources are wasted because of inefficient or non-existent quality management procedures.

When a consensus has been reached on the requirements of a quality management system, it is possible to define a series of generic standards applicable to any type of organisation. These international standards, generically called ISO 9000, are the most widespread, and are generally accepted in all developed countries. The most recent version of the family of ISO 9000 standards consists of four basic interdependent standards which are complemented by other documents, such as guides, technical reports or technical specifications (www.iso.org):

- ISO 9000 (2005): Quality management systems. Fundamentals and vocabulary
- ISO 9001 (2008): Quality management systems. Requirements
- ISO 9004 (2009): Quality management systems. Guidelines for performance improvements
- ISO 19011 (2011): Guidelines for quality and/or environmental management systems auditing.

Demonstrating that a company complies with ISO standards means that its processes have been documented and that the company is systematically auditing and being audited; furthermore, they are following the policies and procedures necessary to produce high-quality products (Arditi and Gunaydin, 1997). Strictly, companies can only be certified under the requirements of the ISO 9001 standard. It is a standard that can be used to certify the efficiency of a quality management system. Nevertheless, if the aim is to improve efficiency, the objectives of the ISO 9004 standard are broader in scope. The principles that underlie the management of quality in these standards are: customer focus, leadership, involvement of people, process approach, system approach to management, continual improvement, factual approach to decision-making and mutually beneficial supplier relationships.

The ISO 9001 standard specifies requirements for a quality management system where an organisation needs to demonstrate its ability to consistently provide products that meet customer and applicable regulations requirements (Hoyle, 2009). It also aims to enhance customer satisfaction through the effective application of the system, including processes for continuous improvement of the system and the assurance of conformity to applicable regulatory requirements. These regulatory requirements focus on the quality management system, management responsibility, resources management, product realisation and measurement, analysis and improvement.

When a quality system is applied to an environment as complex and unique as a construction site, a specific quality plan must be drafted; it is based on applying the company's global system to the specific construction site. The plan must be drafted by the contractor before the start of the construction

> **Box 8.1   Advantages of ISO 9000 quality management systems in the construction industry**
>
> - enhances the company image
> - clarifies definitions of responsibilities in the company
> - improves communication with customers
> - improves company's operating procedures
> - gets better control on subcontractor firms
> - increases customer satisfaction
> - gets better communication among the company's employees
> - increases productivity.

works and will be reviewed throughout its execution. The quality plan is applicable to the materials, activities and services that have been specifically chosen by the construction company in order to comply with the quality requirements stipulated in the contract. The quality plan is drafted for the construction project when a preventive strategy is needed to guarantee quality, even though there might also be a manual in compliance with the ISO 9001 standard requirements. Some advantages of the application of the ISO 9000 set of standards to the construction industry are highlighted in Box 8.1 (adapted from Turk, 2006).

The construction company determines the need to prepare execution documents, work instructions, points of inspection, process files, action plans and so on. for the execution and control of processes, depending on the complexity of the activity, the qualifications of the personnel and the experience of the team (Turk, 2006). The plan establishes the resources required (organisational chart and allocation of human resources) and associated documents (lists, purchasing documentation, machinery, equipment, etc.). The control activities include: verification of compliance with specifications, validation of specific processes, monitoring of activities, inspections and tests. These activities can be defined through the inspection points, testing plans, action plans and, where applicable, specific tests (e.g. load tests for structures).

8.5   Innovation in construction projects

There is no generally accepted definition of innovation at the present time (Baregheh et al., 2009). Innovation is a wide concept that includes such heterogeneous aspects as improvements in processes, products or services (Gibbons et al., 1994). It basically involves incorporating new ideas which generate changes that help solve the needs of a company so that it can increase its competitiveness and improve its position in the market. Similarly, innovation management can be seen as the generation of the suitable circumstances in an organisation to carry out technological, market and organisational changes with a particular degree of uncertainty (Tidd and Bessant, 2009). Fong and Kwok (2009) indicate that innovation is now considered to be a key factor in the success of organisations and creative ideas are seen as a secure parameter for the competitiveness of a firm. Thus, innovation in construction companies can lead to competitive advantages in a demanding and globalised market. This market requires the construction of facilities capable of satisfying the requirements of all the parties involved, taking into account the environment and including the future generations.

Nevertheless, the application of innovation to the construction industry is not a simple task, despite the importance of this sector in the development and growth of any country. Each construction project is different, which means that construction companies have to adapt their processes and resources to each project. Each construction site constitutes a singular prototype whose configuration changes over time. Construction projects are located in different places, and involve the constant movement of personnel and machinery. In addition, the weather and other factors prevent engineers from applying the experiences obtained in other sectors to construction projects (Nam and Tatum, 1988).

Construction companies solve problems of significant technical complexity. For this reason, the technical support department of the company often proposes innovations to specific problems that arise during the execution of the construction works (Yepes et al., 2010). In some cases, they are the result of adopting and adapting ideas from suppliers or from other industries. Although these solutions to specific problems are included in the experience and good practices of the company, innovation is undertaken on a one-off basis, and does not benefit the company nearly as much as if it were incorporated into the organisation's standard management processes.

In any case, innovation should be understood as a systematic and deliberate process, one in which the degree to which the company is connected to the environment plays a very important role (Pellicer et al., 2012). It does not require a complex structure to be successful, but instead should focus on a specific application with the aim of placing the company in a leading position. Innovation changes from timely application of good ideas to a process that can be managed measured and controlled systematically. Consequently, the standardisation of innovation processes constitutes a very important starting point for construction companies.

The key lies in considering innovation as a management process within the company. Therefore, if any process can be standardised – and innovation is considered to be a process – then it can be standardised (Pellicer et al., 2012). Any standards for regulating innovation process management should include the framework of reference, criteria and tools for the identification, drafting and systematisation of each of the activities involved. Under these conditions, each organisation could control and improve the different aspects of innovation and integrate them into the rest of the company's processes. A company that has a standardised management of innovation may collect the following rewards:

- improvement in organisation activities
- improvement in the company's competitiveness in the medium and long term
- better integration of the company's management processes within the company's strategy
- efficient exploitation of the organisation's knowledge
- systematisation of new process and product knowledge
- satisfaction of the future expectations of clients.

There are two families of standards that focus on the standardisation of the innovation process: the British BS 70000-1 standards and the Spanish UNE 166000 standards (Pellicer et al., 2008). BS 70000-1 standards ('Design of innovation systems: guide for the management of innovation') are used as a guide for the development of innovative and competitive products that satisfy the future needs of users. These standards are defined by three features: their purpose is the design of products; they provide a structure for the management (not systematic) of innovation; and they are supported by the quality management ISO 9001 standards.

The Spanish UNE 166000 standards on innovation management consider innovation as a process that can be systematised in a company, using a model similar to the one used for quality or environmental management (Pellicer et al., 2008). Its objectives are to homogenise the criteria, promote the transfer of technology and provide instruments that enable the public administration to assess innovation projects. They also aim to provide companies that have obtained the ISO 9001 certification with an active tool that facilitates the continuous improvement of their processes by means of innovation activities.

Therefore, innovation in the construction industry can be standardised, provided that it is treated as a process. The standard UNE 166002 establishes the basis for the systematisation of innovation in companies (Pellicer et al., 2008); Box 8.2 details the critical success factors for the implementation of an innovation strategy in a construction company. The standard UNE 166002 is process-based, using the methodology 'plan-do-check-act'. This process consists of the following stages (Yepes et al., 2010):

- technological watch: identification of the need and opportunity for innovation by analysing the construction methods during the planning phase, potential alternatives and innovating ideas that will help attain project and company objectives. This stage is heavily influenced by factors such as the scope, complexity and difficulty of the project, market demand, business opportunities, legislation and access to new technologies
- creativity (Figure 8.2): selection of innovation projects at the construction site, at the organisation or due to technological watch – the decision on

> **Box 8.2   Critical success factors for the implementation of an innovation strategy in a construction company**
>
> - construction companies generally innovate in processes
> - the implementation of an innovation management system improves knowledge management
> - construction companies that adopt an innovation management system understand their environment better
> - the control of internal processes (mainly production and management) constitutes a basic source for generating innovative ideas
> - the existence of a quality system certified by the ISO 9001 standard facilitates the implementation of an innovation management system
> - the existence of innovation management system stimulates subcontracting to specialised companies and adds value to the innovation process
> - the active involvement of the site manager in the innovation process has a significant impact on the results of innovation
> - innovation in construction requires the participation of multidisciplinary teams.

innovation projects depends on factors such as the objectives, benefits or competitive advantages expected by the organisation and transfer of innovations to other projects,. The assessment of innovation alternatives must take all project and company objectives into account

- planning and executing innovation projects (Figure 8.2): implementation of innovation projects at the construction site (or in the organisation) – the incorporation of a technological or organisational advance requires the commitment of the whole organisation, the innovation team and the construction site team. The company must allocate the human and material resources needed to carry out the innovation project. This stage is essential, since it involves adjusting scheduled activities to the actual situation

- assessment: the team and the company must assess whether the innovation project's objectives are fulfilled. All the stages of the innovation process should be considered, in addition to any related aspects

- technologic transfer and protection of results: in order for the results to be exploited they need to be successfully transferred to other construction projects. In other words, the innovation process terminates when knowledge is identified, encoded and applied to future projects.

Governments play a prominent role in promoting innovation in the construction industry (Pellicer et al., 2008). They act as promoters, legislators and sources of funding for other organisations. Box 8.3 explains an example of this kind of action in Spain. Within this scenario, a case study of a Spanish contractor is presented in Box 8.4 (based on Pellicer et al., 2012).

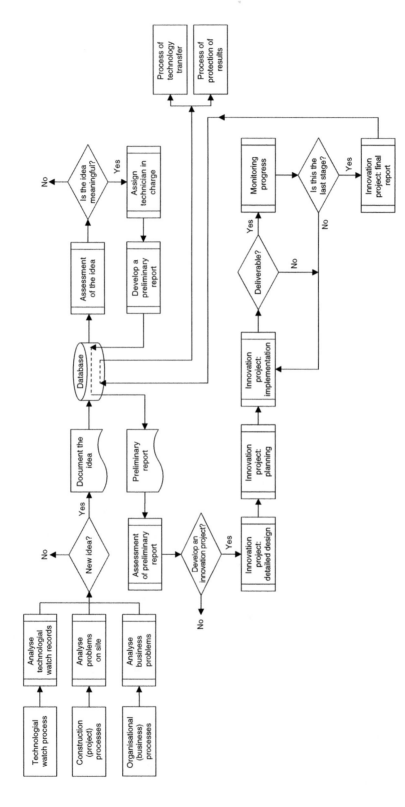

Figure 8.2 Processes of creativity and planning and executing innovation projects (developed from Yepes et al., 2010).

Box 8.3   A case from Spain: the road to innovation through incentives

To bring innovation to markets, governments need support instruments to encourage private investment in R&D. In recent years, bringing into practice some of the commitments adopted by the European Union in Lisbon in 2000, Spain launched two types of incentives to bridge the gap of R&D with other western economies. First, enterprises that invest in R&D activities can obtain tax benefits such as those established in Spanish Law 4/2004 on Income Tax. Second, innovation processes are standardised through the set of standards UNE 166000. These standards, published in 2006, address two fundamental aspects of innovation: the specific design and implementation of projects via the standard UNE 166001, and the management within the company. This process can be standardised following a model similar to those of quality or environmental management via the UNE 166002 standard. Also, at the end of 2006, the Spanish Ministry of Infrastructure began to evaluate how contractors innovate; this Ministry is rewarding construction companies in the bidding process if they show proof of their R&D activities, raising the final score by 10–25%.

Box 8.4   A case study of innovation improvement

A Spanish contractor started its road to innovation, taking into consideration the scenario described in Box 8.3. For three years, it implemented an innovation management system based on the UNE 166002 standard. The performance of the company improved considering various indicators of the overall organisation (revenues, profit before taxes and employees with a university degree) and of the innovation department (employees working in this department and innovation projects with external certification) as displayed below.

Indicator (data from 31 December)	2006	2007	2008	2009	2010	Variation 2009/2007
Revenues (in millions of euros)	451.3	488.1	567.6	591.2	475.6	21%
Profit before taxes (in millions of euros)	6.7	8.2	15.4	17.9	12.2	118%
Employees with university degree	37	42	53	56	48	33%
Employees working at the innovation department	1	3	4	4	4	33%
Innovation projects with external certification (per year)	0	1	1	6	11	500%

(Republished with permission of the American Society For Engineering Management, from Pellicer, E., Correa, C.L., Yepes, V., & Alarcón, L.F. (2012). Engineering Management Journal, 24(2), 40–53.

(Continued)

The certification of the system took place in 2008, at the same time as the economic crisis hit the Spanish construction industry deeply. Examining the data from 2007 to 2009, it can be seen that the employees in the innovation department increased by 33% in order to carry out and certify eight innovation projects. During that period, the employees with university degrees in the whole organisation increased by 33% too, while the profit multiplied six times the value of the revenues.

8.6   Knowledge management in construction

The success, or even the survival, of any organisation depends on how effectively it manages knowledge. Companies have come to understand that knowledge is a resource and a vital asset for the carrying out of their activities, and have come up with various ways of capturing, storing, transmitting and reusing it. Knowledge management has been defined in different ways in the literature. Gurteen (1998) defined this topic as an emerging set of organisational design and operational principles, processes, organisational structures, applications and technologies that helps knowledge workers dramatically leverage their creativity and ability to deliver business value. Whatever the case, knowledge management involves using knowledge in order to increase companies' competitiveness, intellectual capital and the intangible assets of the organisation.

Unlike individual knowledge creation, organisational knowledge creation occurs when the knowledge is managed. A learning organisation is one with the capacity to create, capture and transfer knowledge, but which is also able to modify its behaviour to reflect this new knowledge and experiences (Garvin, 1993). This means the creation of knowledge at all levels and in all areas of the business and that the human element continues to be the most important factor (Carrillo et al., 2004). However, in order for the system to function correctly, there must be a basic level of information and communication technology, which is adapted to the needs of the organisation. The firm must ensure that the correct information is provided to the right person at the right moment in order for them to be able to make the best decision. Thus, a company manages knowledge efficiently when it is able to apply and use knowledge, exploit and explore its resources, adapt to and change its environment and ascertain and develop what it has learned so as to be able to transform it into new knowledge (Petrash, 1996). Finally, in order to innovate, it is necessary to capture external knowledge (technological watch) and generate new knowledge by solving problems on-site and implement solutions to increase the competitiveness of the company (Pellicer et al., 2012).

To sum up, knowledge management is made up by a set of eight factors which must work in harmony with each other in order to organisations efficiently manage knowledge (Castro et al., 2012): knowledge culture, human factors, quality of information, generation of knowledge, knowledge transfer,

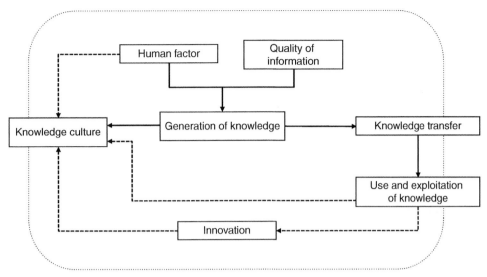

Figure 8.3   The knowledge management cycle in organisations (Castro et al., 2012) (Reproduced with permission from Pontificia Universidad Católica de Chile).

use and exploitation of knowledge, innovation and information and communication technologies. These factors are displayed in Figure 8.3.

Egbu (2004) indicated that, within construction, knowledge is an important resource for organisations due to its ability to provide market leverage and to contribute to organisational innovations and project success. However, due to the fragmented nature of the construction sector, it lacks a coordinated strategy for knowledge management. One of the big problems faced by businesses in the sector is that a lot of their knowledge is held by the professional and technical staff that works on each of its projects. High staff turnover means that good practices are lost and that there is no clear culture that values their capture and management. The efficient management of knowledge would allow construction companies to transmit knowledge across various projects, create synergies inside the organisation, learn from the mistakes and successes of others and receive benefits in terms of productivity and performance.

Construction professionals should consider their past experience as an asset, with an open perspective, which allows them to incorporate new ideas to improve the quality and productivity of their activities, as well as innovate more rapidly than their competitors. As pointed out by Ferrada and Serpell (2009), this is not possible in the construction sector without appropriate knowledge management, where significant efforts have been made to develop and implement systems to manage the capturing, storing and retrieval of explicit project-related information. Face-to-face and group meetings are the principal means utilised in the construction sector for the transfer of knowledge. Ways of acquiring knowledge through job rotation and the capture of knowledge by experienced personnel must be promoted. In order to ensure the quality of the information, it should be selected and organised prior to being stored. A high percentage of the literature describes what businesses are doing or what they think it is best to do when it comes to knowledge management.

Since organisational culture is a prerequisite for the success of knowledge management, a culture of innovation should be promoted. Those cultural and social barriers that make knowledge difficult to manage in the construction sector must be broken down. However, there is still a lot of ground to cover with regard to obtaining better information with regard to how to carry out knowledge management in construction companies.

8.7   Standards and procedures

When the aim is to guarantee the uniformity of a system, process or product, reference patterns are established in documents that are called standards or norms. Compliance with the standard, usually voluntary, is required to unify criteria and facilitate the use of a common terminology in a specific field of activity. The general objectives of standardisation are (Verman, 1973): simplification, communication between the parties involved, production economy, safety and health, protection of consumer interests and the removal of trade barriers.

In any type of company, the set of tasks carried out is so complex that they have to be written down to ensure internal consistency, to preserve them and to make sure that they are methodically applied. These documents are called procedures, and they describe the way in which an activity or process must be carried out.

Standards establish the requirements or performance of products or processes. Procedures are documents drawn up by the company itself and take into account the requirements established in the standards. These documents must include the purpose of the procedure, references to other documents, scope, method and sequence of the tests, acceptance and rejection criteria, key control points and time of inspection. In all cases, the control of a procedure should be documented in the quality records and filed in the quality log at the construction site.

Technical or administrative procedures can be part of a quality management system as they complement the quality manual. In this case, the manual provides a generic description of the company's quality system, while procedures, whether general or specific, establish what is required to attain the objectives listed in the manual (Hoyle, 2009). Procedures must link the ISO standards' requirements and the habitual activities of the company which are considered key to obtaining the required levels of quality. They should include the people involved, information about materials and equipment and a description of the process's key activities. Each organisation should decide which processes should be documented on the basis of customer and applicable legal or regulatory requirements, the nature of its activities and its global corporate strategy.

8.8   Certificates and technical approvals

The quality control of a product or process can be replaced at times with certification of the quality characteristics established by contract or by standards using independent and reliable third parties. The procurement of quality certificates that demonstrate compliance with the standard, enhances the construction work's quality control management. Products that have received officially recognised quality marks are entitled to special treatment; they are exempt from the compulsory supply material controls and reception tests, increasing batch size and improving safety systems (Atkinson, 1995). However, the scope and aim of these quality marks are variable, and an in-depth understanding of the certificates is required to know what they mean. Below are different types of quality certificates, starting with the least reliable (Atkinson, 1995):

- certificate of origin: in this case the manufacturer states that the product complies with some specifications. Although the certificate might not be very reliable, failure to comply with the specifications can be legally actionable. These certificates should have the test results attached
- accredited laboratory test certificate: the test is performed on a small sample, and therefore cannot guarantee the total and continuing production. These certificates should be used with caution due to their small scope
- product certification (type approval): this approves a prototype and therefore does not guarantee the quality of the subsequent manufacturing process
- certification mark: its scope includes continuing production and therefore it is more reliable than the other certificates. When the product is very new and there is no specific standard to regulate it, the certificate is issued in the form of technical suitability documentation.

The European Union Directive 89/196/EEC establishes the approximation of laws, regulations and administrative provisions of the member countries to construction products. The purpose of the directive is to guarantee the free movement of all construction products all over the European Union, through the harmonisation of national laws which regulate the health, safety and welfare essential requirements of these products. These requirements can take the form of harmonised European standard adopted by European standardisation bodies (CEN or CENELEC), or European technical suitability documents if there is no harmonised norm, national norm or European norm mandate. Under this Directive, construction products must have the CE mark, whereby the manufacturer declares that the product complies with the provisions of the European Union Directives. This mark indicates that the product complies with the essential requirements of harmonised norms (EN) and the Guides for European Technical Approval.

Nevertheless, all the countries within the European Union have their own particular set of conditions that have a direct impact on construction (weather, local construction procedures, etc.) and which are not included in the CE mark guidelines. So although the mark facilitates the movement of construction materials between countries, it does not mean that the quality controls established for particular conditions are abolished. This could be solved by the adoption of voluntary norm conformity certificates for each specific case.

When the construction materials and systems are very new (not traditional), the European Organisation for Technical Approvals (EOTA), an umbrella organisation for national authorisation bodies, may draft European technical suitability document guides for a construction product or family of products, acting on a mandate from the Commission (Sedlacek and Müller, 2006). When there is no European standard or European technical suitability document available, products can be assessed and marketed in accordance with existing national provisions and in conformity with essential requirements.

Box 8.5 summarises the system of Assessment and Verification of Constancy of Performance (AVCP) that defines the degree of involvement of third parties in assessing the conformity of the product according to relevant technical specifications (adapted from the Construction Products Association, CPA, 2012).

Box 8.5   Systems of assessment and verification of constancy of performance

To achieve the goal of the AVCP, the Construction Products Regulation (CPR) uses five elements:

1. Factory Production Control (FPC) on the basis of documented, permanent and internal control of production in a factory.
2. Initial inspection of the manufacturing plant and of FPC.
3. Continuous surveillance, assessment and evaluation of FPC.
4. Determination of product type on the basis of the type testing, type calculation, tabulated values or descriptive documentation of the product.
5. Audit testing of samples taken before placing the product on the market.

The five systems of AVCP are explained as follows:

System 1+   product certification comprising the issuing of a certificate of constancy of performance with determination of the product-type, continuous surveillance and audit testing by a notified product certification body

System 1   product certification comprising the issuing of a constancy of performance with determination of the product-type and continuous surveillance by a notified product certification body

---

(Continued)

System 2+	factory production control certification with continuous surveillance by a notified factory production control certification body
System 3	determination of product type by a notified testing laboratory
System 4	manufacturer's task only.

References

Al-Tmeemy, S.M.H.M., Abdul-Rahman, H. and Harun, Z. (2012). Contractors' perception of the use of costs of quality system in Malaysian building construction projects. *International Journal of Project Management*, 30(7), 827–838.

Arditi, D. and Gunaydin, H.M. (1997). Total quality management in the construction process. *International Journal of Project Management*, 15(4), 235–243.

Atkinson, G. (1995). *Construction Quality and Quality Standards. The European perspective*. E & FN, London.

Baregheh, A., Rowley, J. and Sambrook, S. (2009). Towards a multidisciplinary definition of innovation. *Management Decision*, 47(8), 1323–1339.

Burati, J.L., Farrington, J.J. and Ledbetter, W.B. (1992). Causes of quality deviations in design and construction. *Journal of Construction Engineering and Management*, 118(1), 34–49.

Carrillo, P., Robinson, H., Al-Ghassani, A. and Anumba, C. (2004). Knowledge management in UK construction: strategies, resources and barriers. *Project Management Journal*, 35(1), 46–56.

Castro, A.L., Yepes, V., Pellicer, E. and Cuellar, A.J. (2012). Knowledge management in the construction industry: state of the art and trends in research. *Revista de la Construcción*, 11(2), 62–73.

Chan, L.K. and Wu, M.L. (2002). Quality function deployment: a literature review. *European Journal of Operational Research*, 143(3), 463–497.

CPA (2012). *Guidance Note on the Construction Products Regulation (Version 1)*. Construction Products Association, retrieved 02/07/2012 from www.construction products.org.uk.

Dave, B. & Koskela, L. (2009). Collaborative knowledge management: a construction case study. *Automation in Construction*, 18(7), 894–902.

Egbu, C.O. (2004). Managing knowledge and intellectual capital for improved organisational innovations in the construction industry: an examination of critical success factors. *Engineering, Construction and Architectural Management*, 11(5), 301–315.

Ferrada, X. and Serpell, A. (2009). Knowledge management and the construction industry. *Revista de la Construcción*, 8(1), 46–58.

Fong, P.S.W. and Kwok, C. (2009). Organisational culture and knowledge management success at project and organisational levels in contracting firms. *Journal of Construction Engineering and Management*, 135(12),1348–1356.

Garvin, D.A. (1993). Building a learning organisation. *Harvard Business Review*, 71(4), 78–91.

Gibbons, M., Limoges, C., Nowotny, H., Schwartzman, S., Scott, P. and Trow, M. (1994). *The New Production of Knowledge: the Dynamics of Science and Research in Contemporary Societies*, Sage Pub., London.

Gurteen, D. (1998). Knowledge, creativity and innovation. *Journal of Knowledge Management*, 2(1), 5–13.

Hoyle, D. (2009). *ISO 9000 Quality Systems Handbook. Using the Standards as a Framework for Business Improvement* (6th Edition). Elsevier, Amsterdam.

Juran, J.M. and De Feo, J.A. (2010). *Juran's Quality Handbook* (6th Edition). McGraw-Hill, New York.

McCrary, S.W., Smith, R.R. and Callahan, R.N. (2006). Comparative analysis between manufacturing and construction enterprises on the use of formalised quality management systems. *Journal of Industrial Technology*, 22(3):1–8.

Nam, C.H. and Tatum, C.B. (1988). Major characteristics of constructed products and resulting limitations of construction technology. *Construction Management and Economics*, 6(2), 133–148.

Oliver, J. (2008). Knowledge management practices to support continuous improvement. *Journal of Knowledge Management Practice*, 9(4), retrieved 21/10/2012 from http://hdl.handle.net/1959.3/51593.

Pellicer, E., Correa, C.L., Yepes, V. and Alarcón, L.F. (2012). Organisational improvement through standardization of the innovation process in construction firms. *Engineering Management Journal*, 24(2), 40–53.

Pellicer, E., Yepes, V., Correa, C.L. and Martínez, G. (2008). Enhancing R&D&i through standardization and certification: the case of the Spanish construction industry, *Revista Ingeniería de Construcción*, 23(2), 112–121.

Petrash, G. (1996). Dow's journey to a knowledge value management culture. *European Management Journal*, 14(4), 365–373.

Schilling, E.G. and Neubauer, D.V. (2009). *Acceptance Sampling in Quality Control* (2nd Edition). Taylor & Francis Group, Boca Raton.

Sedlacek, G. and Müller, C. (2006). The European standard family and its basis. *Journal of Constructional Steel Research*, 63(11), 1047–1059.

Stuart, M., Mullins, E. and Drew, E. (1996). Statistical quality control and improvement. *European Journal of Operational Research*, 88(2), 203–214.

Tidd, J. and Bessant, J. (2009). *Managing Innovation: Integration Technological, Market and Organisational Change* (4th Edition). Wiley, Oxford.

Turk, A.M. (2006). ISO 9000 in construction: an examination of its application in Turkey. *Building and Environment*, 41(4), 501–511.

Verman, L.C. (1973). *Standardisation: A New Discipline*. The Shoe String. Press Inc., Hamden, (Connecticut).

Yepes, V., Pellicer, E., Correa, C.L. and Alarcón, L.F. (2010). Implementation of a system for achieving innovation opportunities in a construction company. In: *Proceedings of 18th CIB World Building Congress, CIB 2010* (Barret, P., Amaratunga, D., Haigh, R., Keraminiyage, K. and Pathirage C. eds.), 319–330, Salford Quays (United Kingdom).

Further reading

Anumba, C.J., Egbu, C. and Carrillo, P. (2005). *Knowledge Management in Construction*. Wiley-Blackwell, Oxford.

Juran, J.M. and De Feo, J.A. (2010). *Juran's Quality Handbook* (6th Edition). McGraw-Hill, New York.

Tidd, J. and Bessant, J. (2009). *Managing Innovation: Integration Technological, Market and Organisational Change* (4th Edition). Wiley, Oxford.

9 Health and Safety Management

9.1 Educational outcomes

Construction is characterised by its high accident rate, generating a very large proportion of the serious and fatal accidents among the different economic sectors. This is a huge economic and social problem for modern societies. Therefore, a culture of construction safety must be promoted. Following this line of thought, by the end of this chapter, readers will be able to:

- understand the importance of health and safety management, both within the construction company and on the site
- introduce guidelines, methods, documents and criteria, which govern the behaviour of the parties involved in construction regarding health and safety
- analyse the importance of the plan of health and safety on site
- underline the treatment of incidents and accidents as part of health and safety management.

9.2 Introduction to occupational health and safety

Occupational health and safety in construction must be focused on two different areas, which are closely and necessarily interlinked: first, the company in general and, second, every one of the projects that are carrying out that company (Jaselskis et al., 1996). In the first case, the general criteria applied to all projects awarded to the construction company are set, including the prevention procedures applicable to its central offices, facilities, workshops,

Construction Management, First Edition. Eugenio Pellicer, Víctor Yepes, José C. Teixeira, Helder P. Moura and Joaquín Catalá.
© 2014 John Wiley & Sons, Ltd. Published 2014 by John Wiley & Sons, Ltd.

warehouses, construction sites and so on. Nonetheless, the facilities are built in a socioeconomic environment which is subject to legal regulations. Each construction company implements its own methods and procedures, which are applied to every construction site.

This chapter explains the risk–accident cycle in its first part. Later, it describes the legal and business contexts (sections 9.4–6). The management of health and safety in construction projects from the point of view of responsibilities of the parties, planning and control of occupational risks at the construction phase of the facility life-cycle, and treatment of incidents and accidents are dealt with in sections 9.7–10.

9.3   The risk–accident cycle

Occupational health and safety is a process that follows several steps (Turner, 2008; Winch, 2010): identification, analysis, response and control. In this same way, the OHSAS 18001 standard (BSI, 1999) proposes a cycle based on continuous improvement differentiating five steps: establish corporate policies, plan, implement, check, review and improve. These steps are compatible with those of the ISO 9001 standard on quality management system (ISO, 2000). With this background as reference, a five-step 'risk–accident cycle' can be considered (Pellicer and Molenaar, 2009): regulation, education and training, risk assessment, risk prevention and accident analysis (Figure 9.1).

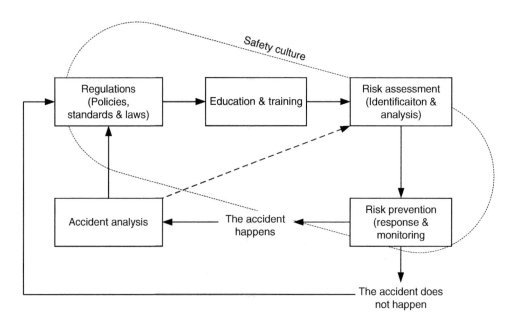

Figure 9.1   Risk-accident cycle (Pellicer and Molenaar, 2009) (With permission from the American Society of Civil Engineers).

Risk assessment comprises risk identification and analysis, as stated in traditional risk management literature. Then risk prevention consists of response and control. In order to stress the importance of setting objectives and organisational learning over time, two previous steps and a final one are added. Regulation is included to emphasise the significance of corporate policies issued by companies, on the one hand, and laws, rules and standards issued by public agencies, on the other. The training and education step reflects the impact that regulations have on people involved, if improvements are to be achieved. Finally, accident analysis is needed to investigate the reason for accidents; in this way, lessons can be learned and other accidents avoided in the future. Obviously, this step is missed if no accident happens.

A break in the loop can appear if regulations (either from the company itself or from public agencies) are not analysed, improved or, at least, implemented; and, similarly, if education and training are not provided. A company that does not apply seriously an occupational safety and health management system, may enter into a spiral of additional risk, making the cycle shorter each time until a serious accident occurs (Pellicer and Molenaar, 2009). But a construction safety culture must be pursued; according to Molenaar et al., (2002) the construction safety culture comprises the knowledge, habits and behaviours that drive the application of safety and health practices, approaches and procedures in the construction industry.

9.4   Regulatory context

The company should establish preventive principles by way of a public statement (Jaselskis et al., 1996). This is used as a basis for corporate policy on the prevention of occupational risks and the development of the necessary actions to carry out the corresponding commitments. These principles must be general and then be developed later into more specific preventive actions.

The mandatory legal regulations for the entire European Union are issued in the form of directives, which apply directly or indirectly to each country's regulations. The most relevant directives that affect the prevention of occupational risks in construction are (https://osha.europa.eu/en):

- 89/391/EEC on the introduction of measures to encourage improvements in the safety and health at work of workers (also known as 'Framework Directive')
- 92/57/EEC on the implementation of minimum safety and health requirements at temporary or mobile constructions sites
- 2009/104/EC concerning the minimum safety and health requirements for the use of work equipment
- 91/383/EEC on supplementing the measures to encourage improvements in the safety and health at work of those workers having a fixed duration or temporary employment relationship

- 94/33/EC on the protection of young people at work
- 92/85/EEC on the introduction of measure to encourage improvements in the safety and health at work of pregnant workers and workers who have recently given birth or are breast feeding.

Of these directives, the 'Framework Directive' (89/391/EEC) establishes general instructions with regard to occupational health and safety in any workplace, company or sector. Directive 92/57/EEC specifically applies to construction projects. Obviously, these two are the ones that most influence the management of occupational health and safety at construction sites, the first for its general nature and the second for its specificity. Finally, the other four are complementary, referring to work equipment, subcontracting, young people and pregnant women.

With regard to internationally applied regulations and standards, those of the International Labour Organisation (ILO, 2001) and the OHSAS 18001 (BSI, 1999) standards should be highlighted. The ILO guidelines provide a management system based on the traditional quality improvement cycle: planning, execution, follow-up and action. The main characteristics of the ILO standard are the following (ILO, 2001): 1. it is not obligatory; 2. it does not attempt to substitute national laws, regulations or standards; and 3. its application does not require a certification.

However, the OHSAS 18001 standards (BSI, 1999) are a set of voluntary international standards related to the management of health and safety. They constitute a tool to facilitate the integration of health and safety requirements with quality and environmental management requirements (ISO 9000 and ISO 14000 respectively). OHSAS 18001 is applicable to industries and organisations of every type and size, without concerning their geographic, social or cultural origin. The aim of OHSAS 18001 is to establish the requirements of a occupational health and safety management system that enables occupational risks to be identified and evaluated. Also, they attempt to define the organisational structure, functions, responsibilities, activity planning, processes, procedures, resources, records and so on. This enables the occupational health and safety policy and its management system to be developed, implemented, reviewed and maintained. The system can be certified following the steps detailed in Box 9.1 (adapted from BSI, 1999).

Box 9.1   Stages to obtain OHSAS 18001 certification

- planning: process and analyse the documentation
- audit: check the effective implementation of the pre-established requirements
- granting: the certificate is granted by a three-year period
- pursuit: audits are made to verify that the conditions of the certification remain.

> **Box 9.2   Main activities of the health and safety management system**
>
> - definition of a business policy of occupational health and safety
> - identification of occupational risks and the related legal norms
> - establishment of objectives and programmes to assure the continuous improvement
> - verification of the system performance
> - improvement of the system.

OHSAS 18001 has three main objectives (BSI, 1999): 1. to minimise occupational risk to employees and other agents; 2. to improve business performance; and 3. to assist organisations in establishing a responsible business policy. The main activities of the health and safety management system are explained in Box 9.2 (adapted from BSI, 1999).

9.5   Agents involved

Chapter 1 explains that construction is a process with different phases: feasibility, design, construction, operation and dismantlement. Numerous agents are involved in each of these phases; some of which influence greatly the subsequent phase (for example, a designer on the construction project), because these phases are interrelated. To ensure adequate operation of occupational risk prevention activities the following agents are considered (based on Directives 89/391/EEC and 92/57/EEC):

- owner: any individual or legal entity for whom the project is carried out
- designer: technical author of the design project
- project manager (also named resident project representative, facultative director, supervisor or inspector): technician who works on the owner's behalf, responsible of the inspection and control of the construction works
- coordinator for safety and health matters at the project design stage: technician appointed by the owner or the designer during the preparation of the design
- coordinator for safety and health matters at the project execution stage: technician appointed by the owner or the project manager during the project's construction
- employer (or entrepreneur): any individual or legal entity (firm or company) who has a labour relationship with the workers. It may be the main contractor (once the project is granted) or a subcontractor
- subcontractors outsourced by the main contractor (first tier) or other subcontractors (second tier and so on): several authors (Tam et al., 2011;

Yik and Lai, 2008) highlight the differences of approach between the main contractor and the subcontractors regarding safety and health
- construction site manager: technician who represents the contractor, and who leads a team of production managers and foremen in order to execute the infrastructure
- worker: any person contracted by the employer.

Other parties are also involved, including the health and safety inspection office for the corresponding public administrative agency, the company's internal and external prevention services, the company's health and safety committee and the suppliers of preventative elements. The key question is: who should manage the prevention on site? The answer is the construction site manager and his/her staff with the support of the company's preventative resources and the supervision of the health and safety coordinator in the execution phase – whoever manages construction should manage health and safety prevention on site too (Burkart, 2002), independently of the support provided by the coordinator or other parties.

9.6   Business context

As indicated above, every company should establish their prevention policy, based on a statement of principles, translating it to the planning, organisation, execution and control processes (Jaselskis et al., 1996). Based on this, the company should implement its own management system for occupational health and safety. It is a basic tool to ensure that the company's prevention services function correctly, and it is integrated into the company's general management system. This system needs to be organised, documented and duly planned, including both the evaluation of occupational risks and the execution of preventative activities, as well as external and internal controls (Koehn and Datta, 2003). It can be certified through the OHSAS 18001 standard, compatible with the ISO 9001 and ISO 14001 standards. The occupational health and safety management system should be based on a continuous improvement process such as explained in Box 9.3 and displayed in Figure 9.2.

9.7   On-site prevention

Organisation of the construction team is not usually too complicated. The construction site manager is the top manager, and all the other engineers, superintendents, foremen and subcontracted companies are under the site manager's authority. The preventive resources for the construction project (whatever they may be) are also under the control of the site manager (Burkart, 2002). Furthermore, the health and safety coordinator forms part of the project management team that represents the owner during the execution of

Box 9.3   Occupational health and safety management system

- prevention policy:
  - definition of principles
  - assumption of commitments and objectives
  - implementation of continuous improvement
  - integration of prevention activities (definition of functions and responsibilities)
  - demonstration of management interest.
- organisation:
  - occupational health and safety committee (political)
  - prevention service (technical)
  - prevention delegates
  - work groups and meetings.
- risk evaluation:
  - identification
  - assessment
  - implementation
  - periodic revision.
- planning:
  - measures to eliminate or reduce risks
  - worker information, training and participation
  - risk control activities
  - actions against foreseen changes
  - actions against foreseen events.
- execution and coordination:
  - procedure implementation
  - documentation of activities
  - internal and inter-company coordination of activities.
- auditing:
  - evaluation of the efficiency of system components
  - strategic plan
  - programme of improvements to be implemented.

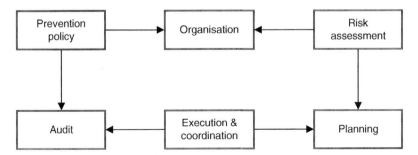

Figure 9.2   Continuous improvement cycle for the occupational safety system (developed from INSHT, 2003).

the project. The most important general prevention principles during the execution phase are (adapted from Reese and Eidson, 2006):

- keep the site in order and at a satisfactory level of cleanliness
- choose the location, taking into account the accesses and particular routes for the transport of equipment and areas for facilities and storage
- indicate the conditions under which the materials are managed
- carry out technical maintenance of facilities and equipment with the aim of correcting any fault, which may affect the workers' health or safety
- establish storage areas for materials, particularly in the case of dangerous substances
- specify the conditions under which dangerous materials are removed, in addition to the storage and removal of debris and waste
- keep in mind their interaction with common on-site activities.

According to the 92/57/EEC directive, the main obligations and responsibilities of each of the agents involved are as stated in Box 9.4.

Box 9.4   Obligations and responsibilities of the agents

- owner (generally acting through the project manager):
  - appoint the health and safety coordinator
  - guarantee that a health and safety plan will be prepared before the construction project is designed and implemented
  - issue basic information to the relevant authorities prior to the beginning of the construction project: this information must be exhibited on site and must contain the date, type of works, owner, project manager, health and safety coordinators (for both phases), contractor, proposed start date, proposed duration, maximum estimated number of workers, proposed number of contractors and freelance workers, and other details regarding the contractor.
- health and safety coordinator during the construction works' execution phase:
  - coordinate general prevention and safety principles
  - coordinate the most relevant statements to ensure that the contractor and the subcontractors apply the prevention principles
  - supervise compliance with the health and safety plan
  - propose and supervise any necessary adjustment to the health and safety plan during the execution phase
  - ensure that the working procedures are applied correctly
  - adopt the necessary measures to ensure that only authorised personnel are allowed to access the site.

(Continued)

- contractor (and subsequently subcontractors), acting through the construction site manager and his/her staff:
  - guarantee the health and safety of the workers
  - adopt the necessary measures to protect the workers' health and safety, including the prevention of occupational risks and the provision of information and training
  - evaluate health and safety risks on the construction site
  - make decisions on the protective measures that should be taken and the protective equipment to be used, where necessary
  - have access to a list of occupational accidents affecting workers and causing them to miss more than three working days
  - prepare reports on occupational accidents for the authorities
  - keep in mind the health and safety coordinator's directions
  - implement at least the minimum health and safety requirements on site
  - consult workers and their representatives, allowing them to participate in discussions regarding health and safety on the construction site
  - ensure that all workers receive adequate training.
- workers:
  - assume responsibilities, insofar as possible, for their own health and safety and that of other people affected by their actions, in accordance with the training and instructions given to them by the employer
  - use machinery, tools, dangerous substances, transport equipment and personal and group protective equipment correctly
  - inform the employer of any occupational situation that represents a serious or immediate danger to safety or any other defect in protective devices.

9.8   Health and safety plan

The health and safety study is mandatory for construction projects in the design phase. Therefore, the design team has to take into consideration the measures needed for occupational risk prevention during the construction phase (Gambatese et al., 2005). Unfortunately, the study often does not fulfil its target for the following reasons:

- the study tends to be carried out in the final phase of the design project; thus, the health and safety coordinator has few options for influencing the design from an occupational health and safety point of view

---

> **Box 9.5   Measures to be included in the health and safety plan**
>
> - risk of entrapment underground, falls, sinking in marshy areas or falls from a height, where the danger is further aggravated because of the nature of the construction work, the processes used or the surrounding working area
> - risks derived from chemical or biological substances which represent a special threat to workers' health and safety or which involve legal requirements to ensure safety
> - ionising radiation, which requires that controlled or supervised areas are designated
> - nearby high voltage lines
> - workers' exposure to drowning risks
> - wells, underground works and tunnels
> - conductors with air supply systems
> - caissons in compressed air environments
> - use of explosives
> - assembly or disassembly of heavy prefabricated components.

- the study tends to be merely adapted from others previously written, without much thought being given to the specific problems of that particular project
- legislation requires that the safety and health plan be an adaptation of the study, so the defects of the study are often transferred into the plan.

Once the procurement stage is over, a contract is awarded to a construction company in order to execute the design project. During the construction phase, a health and safety plan is also mandatory. Directive 92/57/EEC requires that the health and safety plan includes special measures relating to one or more of the categories indicated in Box 9.5.

The plan should include the identification, assessment, implementation and control of prevention measures (Burkart, 2002). A proposal of table of contents for the health and safety plan is included in Box 9.6.

9.9   Management of the health and safety plan

It is essential that all the parties involved in the construction project are conscious of the importance of occupational health and safety. It is also particularly important that the planning of the construction project includes the foreseen risks, their evaluation and the measures to avoid, mitigate or eliminate such risks. Nevertheless, the necessary practical and detailed

Box 9.6   Table of contents for the health and safety plan

- purpose
- background
- objectives
- constraints
- description of the project:
  - activities
  - construction processes
  - layout of the construction site
  - description of the process by phases and activities, detailing description, equipment, additional resources, labour, materials, stock and so on.
- project schedule
- critical activities for prevention
- risk identification during the construction phase for:
  - the site layout
  - every activity, keeping in mind equipment, additional resources, labour, materials and execution
  - special risks
  - concurrence of risks.
- prevention measures:
  - collective protection
  - individual protection
  - signalling risk and safety measures.
- risk assessment
- effectiveness of protection measures
- workers' support facilities: hygiene services, changing rooms, eating areas and so on
- accident support and investigation:
  - first aid
  - prevention medicine
  - evacuation
  - accident and incident investigation.
- training and information for workers and operatives:
  - risk information
  - initial training
  - continuous training
  - working instructions.
- emergency plans
- prevention during the implementation phase:
  - risk identification
  - protection
  - assessment.

(Continued)

- management of health and safety plan:
  - monitoring and control documents
  - changes and modifications
  - safety committee.
- drawings
- technical specifications
- budget.

management procedures to carry out the prevention and control plan are also required (Burkart, 2002).

Analysis of a good sample of current health and safety studies and plans has led to the conclusion that most of them do not include a health and safety management facet. When they do, they usually miss out the timing of the process. In other words, the plan includes the what and who, but not the how and, specially, the when. The answers to these last two questions are essential in order to develop adequate occupational health and safety management. It is necessary that those responsible at the site be aware of the importance of health and safety management.

A health and safety plan that includes all the documents, steps and procedures, allowing compliance with the set objectives, should be used for adequately managing the construction project (Burkart, 2002). Furthermore, from a strategic point of view, the most important activities have to be selected in order to control them properly (Hallowell and Gambatese, 2010); this is a key aspect for the management of health and safety in small and medium-size companies.

In the following paragraphs, two aspects of the health and safety management are considered: the execution of the construction works (Reese and Eidson, 2006) and the information and document flow throughout the project (Civitello and Levy, 2005). An important first step is the generation of standardised guidelines, forms and two-way tables, one for each activity or work stage on site, specifying the areas where the activity is developed, and the other the performance factors and control indicators (from a visual inspection to a complex test), as shown in Figure 9.3. These forms and procedures should be computerised in order to improve management in real time.

The aim of the table is to ensure that all preventive activities and controls are documented (Civitello and Levy, 2005). An additional goal is to verify two facets (Reese and Eidson, 2006): 1. Every activity must be taken into consideration (timbering, assembling, setup of nets and railings, etc.) at different times and stages in the process to be defined ahead of time; and 2. Checkpoints must be implemented in each area of the construction site (for example, at particular intervals in the foundation slab). Within each activity,

ACTIVITY: FOUNDATIONS		Depth	Slope	Shoring	...
Phase or Area A					
Control 1 (date and time)					
...					
Control n (date and time)					
Phase or Area B					
Control 1 (date and time)					
...					
Control n (date and time)					

Figure 9.3   Control of activities form.

WORKERS' TRAINING	Date	Hours	Issuing Party	Documentation	Signature
Activity A					
Worker 1					
...					
Worker n					
Activity B					
Worker 1					
...					
Worker n					

Figure 9.4   Control of workers' training form.

the specific controls to be carried out should be specified (Reese and Eidson, 2006), for example for foundations: depth, appropriate slope, quality of the shoring, distance between evacuation stairways and so on. These tables can be applied to non-production activities, which are nonetheless necessary and mandatory, such as training, ensuring that it is correctly adapted (Figure 9.4).

Even though the possibilities of generating forms are enormous, a specific model for monitoring and control of the construction project is displayed in Figure 9.5. As stated in Chapter 4, one of the goals of these forms is to document effectively what is happening at the site, as well as the information flow among the different stakeholders; forms and procedures explained in Chapter 4 can be implemented specifically from the health and safety viewpoint. Another goal is controlling every activity in the area that is performed during different times or stages defined previously (linked to the project schedule). Each activity should specify the controls that should be performed; for example, in a trench, several performance factors and indicators have to be set in advance in order to control its depth, slope adjustment, quality of shoring and bracing, debris loads, distance between stairways and so on. Not

HEALTH AND SAFETY CONTROL LOG

DAY: REPORT N°:
PROJECT:
CONTRACTOR:
SITE MANAGER:
OWNER:
PROJECT MANAGER:
HEALTH AND SAFETY COORDINATOR:

EXECUTION OF WORKS

TYPE	LOCATION	CORRECT	DEFICIENT	INEXISTENT	INCIDENCES AND OBSERVATIONS
SIGNS, NOTICES AND NOTIFICATIONS					
EXCAVATIONS AND TRENCHES					
SHORING					
SLOPE REPOSE					
LADDERS					
STOCKPILE OF EXCAVATED MATERIAL					
REMOVAL OF EXCAVATED MATERIAL					
BARRICADES & RAILINGS					
TUNNELS					
BLASTING ARRANGEMENTS					
APPROVED SHORING DESIGNS					
EXCAVATIONS PROPERLY DEWATERED					
PROPER VENTILATION					
...					
OPENINGS IN WALLS, FLOORS AND ROOFS					
SCAFFOLDS					
MATERIAL HANDLING					
ELECTRICAL EQUIPMENT					
HAND TOOLS					
HOUSEKEEPING					
FIRST AID					
TRAFFIC CONTROL					
PERSONAL PROTECTIVE EQUIPMENT					
HEAVY EQUIPMENT					
SECURITY					
...					

CONSULTATIONS / MEETINGS:

FROM	TO	SUBJECT	DECISIONS MADE

DOCUMENTATION:

FROM	TO	TYPE	SUBJECT

Signed by...   Distributed to...

Figure 9.5   Example of control form (developed from Civitello and Levy, 2005).

only safety risks have to be considered, but also others related to hygiene (physical, chemical and biological), ergonomic and psychosocial, such as stress and burnout (Lingard and Rowlinson, 2005).

Using these types of tools, the construction site manager or the company's technician in charge of health and safety issues can manage the prevention of construction risks with greater guarantees. In parallel, there is the control implemented by the health and safety coordinator on the owner's behalf. Basically, the management and control of occupational risk prevention in construction should not remain a mere summary of the activities carried out by each party involved; it should also include a set of procedures and documents, which provide reliable proof that the necessary steps designed to ensure adequate prevention of occupational risks on site have been carried out and controlled (Civitello and Levy, 2005). Finally, managing the prevention of occupational risks should foresee the possibility of modifications and changes, so that they can be rapidly adapted to the whole health and safety management process.

As stated previously, one of the purposes of using these procedures, guidelines and forms is to document everything properly. This can be very useful in a number of ways (Reese and Eidson, 2006):

- to analyse the causes of risk scenarios, incidents and accidents in order to diminish the accident rate
- to justify, in the event of an incident or an accident that everything is correct, in order to mitigate sanctions or penalties
- to produce the documentation required by certification standards such as OHSAS 18001.

The simulation of construction processes using BIM or other tools (explained in Chapter 4) could establish what activities have to be checked and when, so that guidelines for performing the work safely could be set up in advance, as well as specific training and monitoring.

Regarding the responsibilities for health and safety at the construction site, the construction site manager is the one in charge. He/she can delegate tasks to one or more specific technicians that are competent in the matter, and dedicated exclusively to perform these tasks. If the project is complex, several health and safety delegates can work together, generally by area or specialty. Furthermore, the role of the coordinator of health and safety during the execution of the works is always essential because it provides a monitoring process from the owner's point of view. In many countries, this agent has legal responsibilities in the case of an accident. Anyway, they have to work as a team and be motivated to reduce risk and enhance health and safety (Mohamed, 2002). A lean construction approach (explained in Chapter 11) would encourage every stakeholder to be responsible and competent of their own job; at the construction site, this can also be implemented using the Last Planner System (see Chapter 12 for details).

Audits for the prevention of occupational risks are the most common method of determining the effectiveness and efficiency of any system (Reese and Eidson, 2006). Reference should be made to what constitutes the inspection of control. It can be internal or external. The audits themselves are aimed at the company as a whole, but in the case of construction companies, they should also clearly include construction projects as the basic production components in these companies.

9.10 Incidents and accidents during construction

The incident log records any complaint or protest, by any of the agents involved, during the construction phase, regarding occupational risks at the site; this impeachment is conveyed to the public administration in charge of labour affairs and occupational inspection issues, which can act consequently applying corrective measures (Lingard and Rowlinson, 2005). This log is kept on site by the construction site manager. The coordinator for health and safety matters can access the log as well as the subcontractors, freelance workers, the workers' representatives and the technicians from specialised administrative bodies on health and safety. All of them can write down complaints in the log.

Once data is entered into the incident log, the coordinator is obliged to submit a copy of the record to the works inspector (representative of the public agency). Furthermore, he/she must communicate this fact to the contractor, subcontractors and workers' legal representatives. If the health and safety coordinator or any other person observes any non-compliance with the health and safety measures during the execution of the construction project, he/she should advise the contractor, leaving a record of this non-compliance in the incident log. Furthermore, the coordinator has the authority to suspend the tasks, or even the project as a whole, in the case of serious or imminent risk to the health and safety of the workers.

If the inspector confirms that the standards on occupational risk prevention are not observed, the inspector may order the immediate suspension of certain tasks or all of the construction works. The project will resume on the inspector's orders, as soon as the causes of these risks are removed. The stoppage of the works in these cases is without prejudice to meeting contract deadlines. That is, there is no excuse for delays in completion of activities or the project itself because of health and safety shutdowns. If an activity is interrupted because of lack of safety, the schedule is not frozen or changed (Lingard and Rowlinson, 2005).

In the case of accidents certain protocols must be followed, and they must be provided in the safety and health plan: first aid, evacuation,

---

ACCIDENT INVESTIGATION REPORT FORM

REPORT N°:	PROJECT:
DAY OF ACCIDENT:	CONTRACTOR:
TIME OF ACCIDENT:	SITE MANAGER:
PLACE OF ACCIDENT:	OWNER:
DATE OF REPORT:	PROJECT MANAGER:
REPORTED BY:	HEALTH AND SAFETY COORDINATOR:

TYPE OF ACCIDENT: (check one)

- **Vehicular**
- **Personal**
- **Other**

INJURED: (repeat as many times as needed)

- **Name**
- **Address**
- **Trade**
- **Insurance Carrier**

DETAILED DESCRIPTION OF THE ACCIDENT: (attach additional documents if needed)

TIME LOST:
- **By the injured worker**
- **By other co-workers (not injured)**

WAS THE RISK DETECTED IN THE H&S PLAN?

WAS A WORK PROCEDURE IMPLEMENTED FOR THE JOB?

WAS SAFETY EQUIPMENT IN USE?

EQUIPMENT OR MATERIALS INVOLVED (if any):

WHAT CAUSED THE ACCIDENT? (the cause/s of the accident can be codified)

PROBABILITY OF RECURRENCE: (low / medium / high)

PERSON IN CHARGE OF THE CORRECTIVE MEASURES:

SEVERITY OF THE ACCIDENT: (low / medium / high)

CORRECTIVE MEASURES PLANNED:

CORRECTIVE MEASURES ALREADY IMPLEMENTED:

Signed by…   Distributed to…

Figure 9.6   **Example of accident investigation report form (developed from Civitello and Levy, 2005).**

protection of the other employees, work stoppage (if needed), investigation of the accident and other additional measures (Reese and Eidson, 2006). Regarding the investigation, a report must be completed whenever there is any noteworthy incident (Civitello and Levy, 2005), even if no workers are injured, as shown in Figure 9.6. The causes of the accident can be codified according to any criteria; generally, each country has regulations in order to provide a base for later exploitation of the data.

References

BSI (1999). *OHSAS 18001: Occupational Health and Safety Management Systems: Specification*. British Standards Institution, London.

Burkart, M.J. (2002). Wouldn't it be nice if… *Journal of Construction Engineering and Management*, 7(2), 61–67.

Civitello, A. and Levy, S. (2005). *Construction Operations Manual of Policies and Procedures*. McGraw-Hill, New York.

Gambatese, J.A., Behm, M. and Hinze, J. (2005). Viability of designing for construction worker safety. *Journal of Construction Engineering and Management*, 131(9), 1029–1036.

Hallowell, M.R. and Gambatese, J.A. (2010). Population and initial validation of a formal model for construction safety risk management. *Journal of Construction Engineering and Management*, 136(9), 981–990.

ILO (2001). *Guidelines on Occupational Safety and Health Management Systems (ILO-OSH 2001)*. International Labour Organisation, Geneva.

INSHT (2003). *Manual de Procedimientos de Prevención de Riesgos Laborales: Guía de Elaboración*. Ministerio de Trabajo y Asuntos Sociales, Madrid (in Spanish).

ISO (2000). *ISO 9001: Quality Management Systems: Requirements*. International Organization for Standardization, Geneva.

Jaselskis, E.J., Anderson, S.D. and Russell, J.S. (1996) Strategies for achieving excellence in construction safety performance. *Journal of Construction Engineering and Management*, 122(1), 61–70.

Koehn, E.E. and Datta, N.K. (2003). Quality, environmental, and health and safety management systems for construction engineering. *Journal of Construction Engineering and Management*, 129(5), 562–569.

Lingard, H. and Rowlinson, S. (2005). *Health and Safety in Construction Project Management*. Spon Press, Oxon (UK).

Mohamed, S. (2002). Safety climate in construction site environments. *Journal of Construction Engineering and Management*, 128(5), 375–384.

Molenaar, K., Brown, H., Caile, S. and Smith, R. (2002). Corporate culture: a study of firms with outstanding construction safety. *Professional Safety*, 47(7), 18–27.

Pellicer, E. and Molenaar, K.R. (2009). Discussion of 'Developing a model of construction safety culture'. *Journal of Management in Engineering*, 25(1), 44–45.

Reese, C.D. and Eidson, J.V. (2006). *Handbook of OSHA Construction Safety and Health* (2nd Edition). CRC Press, Boca Raton (Florida).

Tam, V.W.Y., Shen, L.Y. and Kong, J.S.Y. (2011). Impacts of multi-layer chain subcontracting on project management performance. *International Journal of Project Management*, 29, 108–116.

Turner, J.R. (2008). *The Handbook of Project-Based Management: Leading Strategic Change in Organizations*. McGraw Hill, New York.

Winch, G.M. (2010). *Managing Construction Projects* (2nd Edition). Wiley, Oxford.

Yik, F.W.H. and Lai, J.H.K. (2008). Multilayer subcontracting of specialist works in buildings in Hong Kong. *International Journal of Project Management*, 26, 399–407.

Further reading

Howarth, T. and Watson, P. (2008). *Construction Safety Management*. Wiley, Oxford.

Reese, C.D. and Eidson, J.V. (2006). *Handbook of OSHA Construction Safety and Health* (2nd Edition). CRC Press, Boca Raton (Florida).

10 Environmental and Sustainability Management

10.1 Educational outcomes

The environmental integration of the construction process is a key element that the owner, the designer and, of course, the contractor should always keep in mind. The concept of sustainability of the built environment has already permeated the industry, and new ideas, such as construction and demolition waste or green building, are already being used by the interested parties. Hence, by the end of this chapter readers will be able to:

- understand the concepts of environmental impact assessment and environmental management
- outline the European legislation related to environmental integration issues, and the regulations pertaining to environmental management
- apply environmental management to the construction site
- illustrate the construction and demolition waste, as well as its reduction, reuse and recycling
- identify the BREEAM and LEED certification systems for green buildings.

10.2 Environmental impact assessment

Any human activity, including construction, impacts on the environment, even residually (in some rare circumstances positively). Every ecosystem undergoes changes regardless of human action. The planned facility should keep compatible links with the ecosystem in order to preserve it, give it added

Construction Management, First Edition. Eugenio Pellicer, Víctor Yepes, José C. Teixeira, Helder P. Moura and Joaquín Catalá.
© 2014 John Wiley & Sons, Ltd. Published 2014 by John Wiley & Sons, Ltd.

value as a resource, delay environmental degradation, or even restore the original system by altering the existing conditions. In summary, the facility forms part of the environment. Environmental impact can be defined as any change to the environment whether adverse or beneficial, wholly or partially resulting from an organisation's activities, products or services (ISO, 2004). Therefore, the prediction, assessment, prevention and mitigation of environmental impact on construction projects are undertaken through an environmental impact assessment procedure (Ridgway, 1999). This procedure can take a life-cycle approach, evaluating the environmental effects of the facility in each of the phases of its life-cycle (feasibility, design, execution, operation and dismantlement).

For the feasibility and design phases, the parameters that determine some of the impacts that will occur in the rest of the phases are defined (fundamentally the impact as an 'element implanted' in nature). In this first phase, and as part of the definition of the facility, the materials whose manufacturing or procurement process could cause impacts have to be identified. In other cases, a project which appears to be more 'ecologically appropriate' than others could have more unacceptable impacts (Chen et al., 2005). For example, the construction of a wall looks forward to ensuring that it has a minimum visual impact, even though it may have potential impacts such as that of a biological barrier. This could be achieved with 'green walls' as opposed to the classic concrete structures. Green walls are fully integrated into nature but they require geosynthetic products to build them; these may be composed of irreversible compounds and have very polluting manufacturing processes. The advantage of green walls over concrete structures may be reduced if the contaminants produced when manufacturing and installing geosynthetics and concrete are compared.

In many cases, the greatest negative impact occurs during the construction phase (Gavilan and Bernold, 1994; Tam and Tam, 2006); this phase usually receives the most attention. It is linked to the two previous phases, because the methods and procedures have to be determined a priori to prevent them from remaining outside the concern of the future inspection (Chen et al., 2005); most of this chapter focuses on the environmental assessment considering the interface between the design and construction phases of the facility life-cycle.

During the operation phase, the owner of the facility develops a follow-up strategy to demonstrate that it complies with objectives, regulations and policies (Lundberg, 2011). The monitoring of the operation can detect additional environmental effects; thus, the owner can understand and learn the underlying causes and improve its environmental procedures and actions. Despite the importance of this operation phase, it is actually often neglected, or inadequately treated. Environmental management systems can help to perform this important task during the operation phase of the facility life-cycle.

10.3   Basic legislation for environmental impact assessment

In the construction process, the starting point is satisfying a need or solving a problem. However, a new element appeared in this process during the second half of the 20th century: prevention of damage to the environment and good management of natural resources, was added to the specific needs of each project (van Bueren and de Jong, 2007). In 1969, the United States Environmental Policy Agency (EPA) enacted the National Environmental Policy Act (NEPA) to provide a national policy that would encourage harmony between humankind and the environment, and promote efforts which would prevent or eliminate damage to the environment and the biosphere. Furthermore, the USA enacted other laws in keeping with NEPA, even before Europe started the environmental legislation process. Those are: the Clean Air Act (1970), the Occupational Safety and Health Act (1970), the Endangered Species Act (1973), the Resource Conservation and Recovery Act (1976), the Toxic Substances Control Act (1976) and the Clean Water Act (1977) (Eccleston, 2011).

Europe did not issue any type of regulations until 1985: Directive 85/337/EEC on the assessment of the effects of certain public and private projects on the environment. It was subsequently amended and enlarged by Directives 97/11/EC, 2003/35/EC and 2009/31/EC. These four directives have been revised and unified into Directive 2011/92/EU. The primary directive establishes the general principles of the environmental impact assessment (EIA). Annex I of this directive lists the projects which must be subject to an EIA, while Annex II registers the projects for which each country should undertake an ad hoc study and establish and apply certain restrictions to determine whether the project should be assessed or not (Marsden and Koivurova, 2011). Several additional directives have been published subsequently, the most important being:

- directive 92/43/EEC on the conservation of natural habitats and of wild fauna and flora, amended by Directive 97/62/EC and Regulation EC 1882/2003
- directive 2000/60/EC, a framework for action in the field of water policy
- directive 2001/42/EC on the assessment of the effects of certain plans and programmes on the environment, that all environmental aspects should be included during the feasibility phase
- directive 2003/4/EC on public access to environmental information
- directive 2004/35/EC environmental liability with regard to the prevention and remedying of environmental damage
- directive 2008/50/EC on ambient air quality and cleaner air.

The environmental impact assessment, as defined under European legislation (mainly Directive 2011/92/EU), is a process designed to predict and inform about the effects that a given project may have on the environment. The environmental impact assessment comprises two different parts (Marsden and Koivurova, 2011):

- environmental impact study: it predicts how the execution of a project may impact on the environment and establishes the appropriate corrective measures
- environmental impact statement: this is a report based on the environmental impact assessment process. It is issued by the corresponding environmental agency, after the environmental impact study has been reviewed and any issues raised in the public hearings and the planning stage have been analysed. It examines the feasibility of the project and, if it is approved, establishes the conditions, which must be complied with in order to adequately protect the environment and natural resources. It also includes the procedure for monitoring activities, in line with the environmental monitoring programme.

10.4   Environmental management tools

According to the European Environment Agency (http://www.eea.europa. eu), the environmental tools that small and medium-size enterprises (between 80 and 90% of companies in the European Union) use for environmental management focus on processes or products. From the point of view of construction companies and construction projects, the process-based instruments are the most important ones.

Environmental management takes into account the organisation's activities when they affect, or may affect, the environment. In this sense, it is possible to link the input supply chain of a business production process to the environment. The advantages of adequate environmental management in a company can be summarised as follows (Eccleston, 2011): cost cutting, ensuring compliance with current legislation, anticipating compliance with predicted future legislation, reducing environmental risk, satisfying supply chain requirements, developing relationships with owners and agencies, improving public image, increasing market opportunities and satisfying workers.

Given that a company works as a system that operates with material and energy flows managed by people (while using other socioeconomic resources), the appropriate environmental management of the different elements in the system can optimise and cut costs, reducing waste (Eccleston, 2011). A broader definition of construction waste is associated with the use of resources that do not add value to the construction process or to the built facility, linking this concept to the lean construction philosophy (Formoso et al., 2002) as explained in the Chapter 11.

The ISO 14000 family of standards follows the ISO's action guidelines on the environment (ISO, 2004). These are related to various environmental management instruments and are the basis for regulations whose territorial scope is more limited (e.g. the EMAS regulations in the European Union). The ISO 14001 standard has been in use in every European Union country since 1996. It is a specification standard comprising a set of requirements that need to be established and maintained in an environmental management system. By complying with these requirements, a company can demonstrate to the outside world that it has implemented an appropriate management system. One way in which a company can show that it has met the standards requirements is by performing an internal audit in which the firm verifies its own compliance with the requirements. Nevertheless, the company may feel that they will have more credibility in the marketplace if their compliance with the ISO 14001 standard requirements is verified by independent third parties. This third-party recognition is known as certification. The competence of the certification issuer is vouched for by an accreditation certificate (Marsden and Koivurova, 2011).

The EU Eco-Management and Audit Scheme (EMAS) is a management system for organisations to evaluate their environmental performance (http://ec.europa.eu/environment/emas/). Even though it was originally restricted to manufacturing sectors, since 2001 it has been open to the other economic sectors through Regulation 761/2001 of the European Parliament and Council. This system allows additional compliance with environmental legislation ensured by government supervision, in order to verify its correct implementation in an organisation (Marsden and Koivurova, 2011).

10.5   Environmental management at the construction site

During the design phase, an in-depth analysis has to be performed in order to examine the environmental consequences of the implementation and operation of the facility at the site. This analysis is summarised and written down in the environmental impact study. This study describes the predicted environmental impacts of the project and the mitigation measures that must be taken in order to avoid, reduce or offset the impacts. This environmental impact study is approved by the administrative authority in charge of environmental issues becoming an environmental impact statement (Jain, 2012).

To ensure that the requirements drawn from the environmental impact statement are complied with during the construction phase, the owner must appoint a technician (the environmental manager) who will inspect the construction project to verify that the commitments of the statement are met according to the European regulations stated previously (mainly Directive 2011/92/EU). The environmental manager is part of the project management team. If the owner does not appoint anyone to the post, the project manager can act as a substitute. His/her main obligation is to guarantee compliance with the environmental impact statement and the environmental monitoring programme. The most important aspect of this work is to monitor and control the construction company that is undertaking the construction project, to ensure that they implement the appropriate corrective measures as required, and subsequently to verify that they are functioning correctly. To achieve this, the environmental manager must make sure that the contractor has implemented all the necessary processes and is using an appropriate organisational structure (Chen et al., 2005). The issuing of periodic reports is mandatory. They must also be issued for project milestones (layout or handover, for example) or if unforeseen problems occur. The reports are issued by the environmental manager and sent to the environmental agency in charge of the construction project, which issued the corresponding environmental impact statement (Jain, 2012).

The supervision of the environmental management at the construction site is also the responsibility of the environmental manager, who should keep the project manager informed at all times. The environmental management at the construction site could also form an intrinsic part of the construction company's environmental management system, such as ISO 14001 (Ridgway, 1999). The main objective could be to establish information channels and working relationships between the company and the owner through the project manager. It could include management of the earth-moving process (clearing, vegetation, quarries, extraction, etc), dredging, temporary works, solid waste, contaminant emissions (including noise, vibrations and dust), fires and so on. The appropriate licences and authorisations must be obtained (Jain, 2012).

In important construction projects, due to their size or impact on the surrounding environment, an environmental management team may be formed, which includes experts who prepare a project monitoring plan. For example, archaeologists (effects on historical-artistic assets), biologists (effects on flora and fauna), geologists (effects on soil), landscape architects (effects on landscape), hydrogeologists (effects on underground water), chemists (pollution effects), or physicists (effects on noise and vibrations) are frequently involved, depending on the type of construction project and the area affected (Eccleston, 2011).

10.6   Construction and demolition (C&D) waste management

Construction and demolition (C&D) waste is one of the sub-products that comes from the construction, renovation (including refurbishments and repairs) and demolition of facilities: buildings or civil engineering infrastructures (Figure 1.2 in Chapter 1). These sub-products often contains materials that are heavy and bulky, such as concrete, asphalt, metals, masonry, glass, plastics, timber and salvaged building components (doors, windows and plumbing fixtures); it also includes earth and rock from excavation, clearing sites and dredging (Kofoworola and Gheewala, 2009). It is usually described as the debris or rubble taken out from the construction site and got rid of in landfills. As stated previously, the broader interpretation of waste associated with the lean construction philosophy (Formoso et al., 2002) will not be employed in this chapter.

Even though, in several ways, C&D waste is considered inert, that might not always be the case (Fatta et al., 2003); waste can be also hazardous (toxic, radioactive or infectious). Annex III of the Council Directive 91/689/EEC establishes the criteria for waste hazardousness (later updated by Council Decision 94/904/EEC and Commission Decision 2001/118/EC). Some materials are hazardous: asbestos is probably the most well-known, but also lead, tar, additives and others can be toxic or flammable in some specific circumstances. There are other materials (e.g. paint and plastics) that may not be inert. Some materials can turn out to be hazardous because of the environment or through contact with others (Fatta et al., 2003). The basic definitions can be found in Directive 2008/98/EC on waste.

Commission Decision 2000/532/EC of the European Union (later amended by several Council Decisions, Commission Decisions and Directives) established a list of wastes (Marsden and Koivurova, 2011). The section regarding construction and demolition wastes (including excavated soil from contaminated sites) proposes the following breakdown for the first level:

1. concrete, bricks tiles and ceramics
2. wood, glass and plastic
3. bituminous mixtures, coal tar and tarred products
4. metals (including their alloys)
5. soil (including excavated soil from contaminated sites), stones and dredging spoil
6. insulation materials and asbestos-containing construction materials
7. gypsum-based construction materials
8. other construction and demolition waste.

The increasing investment in construction throughout the past decades, as well as the abandonment of good practices for optimising the use of natural resources used during the first half of the 20th century (see the example in Box 10.1), has greatly increased C&D waste. According to Solis-Guzman et al. (2009), 35% of the waste worldwide comes from C&D. Aldana and Serpell (2012) summarise several local studies, providing data that can highlight the importance of C&D waste in developed and developing economies: 15% in Finland, 20% in Germany, around 25% in the Netherlands and France, 20–30% in Australia, Brazil and the United States, 30% in Canada and Italy, but more than 30% in Chile, China, Japan, Spain and the UK; these figures are approximate values that come from different sources, adding the problem of homogenisation into the definition and exploitation of data, different construction cultures, as well as the difficulty of data procurement (timing and criteria) (Fatta et al., 2003).

The handling, storage and treatment of C&D waste produces serious problems related to its quantity and quality. Regarding the first issue, the

Box 10.1   Optimisation of natural resources using mass diagrams

Until the 1970s, investment in Spanish roads was small. Engineers had to optimise the use of resources. Earthmoving was an important part of the construction project, often accounting for more than 25% of the construction costs. Earthmoving comprises the excavation of the earth material and the placement of the fill material for the embankment. If the earth movement is balanced across the construction site, there is a minimisation of additional material for the embankment as well as hauling of waste material and new filling material coming from external sites and quarries. In order to do that properly, using a manual method (computers were not easily available at that time), every cross-section had to be analysed, calculating the areas (and later the volumes) in cut and the areas in fill. A mass diagram represents graphically the cumulative amount of earthwork moved along the centreline and the distances over which the earth has to be transported; this way, the diagram shows the average haul and overhaul for each section of the road, as well as the quantity of material moved and how far it is moved. Generally, the mass diagram is included in a time-space diagram (an adaptation of a line of balance for roads and railways projects) as depicted in Chapter 12 (Figure 12.11). This design procedure (including the mass diagram) was used by engineers working in the public service for years.

Nevertheless, during the 1980s and 1990s, the Spanish administration invested a huge amount of money that enabled the construction of 15 000 km of freeways in 20 years, approximately, with the help of the European Regional Development Funds. During those two decades, road design was outsourced by the public agencies. Furthermore, the technical specifications for filling materials used in the embankment were more restrictive because of new regulations; thus, higher-quality filling materials were mandatory for meeting the requirements. In most of the projects, this precluded the use of the excavated material for filling the embankment. For this reason, and because of the tight deadlines for design and construction, mass diagrams were abandoned by designers. During those two decades a lot of filling material was extracted and hauled from quarries, while a lot of excavated materials were treated as waste and taken to landfill.

Recent European and Spanish regulations have led to reconsidering the old methods, specially the balancing of earthmoving using the mass diagram. Thus, sustainability is improved in the current and future road design, by trying to utilise the excavation material as filling material for the embankment. Some relaxation in the technical codes has helped too.

amount of waste generated nowadays fills the legal rubbish dump and landfill sites, and triggers the use of new and illegal dumps that damage the environment (Gavilan and Bernold, 1994); thus, there is a need to set a limit of C&D that can be dumped at an authorised site. Furthermore, the kind of waste generated is also a key issue; as stated, hazardous materials (on its own and in combination), even in small quantities, can affect the health and safety of the operatives involved in all of the processes, as well as impacting the environment dramatically, possibly in an irreversible way. If these C&D hazardous materials are treated, stored or disposed of in an uncontrolled way, its potential for contamination represents a menace to nature, in general, and to human beings, in particular (Pongrácz and Pohjola, 2004).

There are two strategies we can use to minimise the amount of waste generated through the construction process (McDonald and Simthers, 1998): 1. Employ source reduction techniques during the design, procurement and construction stages of the process; 2. Optimise the management of waste. Regarding the latter, Johnston and Mincks (1995) anticipated that a right waste management strategy could be not only beneficial, but also profitable for the contractor.

Because of all these problems, a waste management strategy based on recycling and reuse has been encouraged since the end of the last century. For example, the European Union, within the Sixth Environment Action Programme and Directive 2008/98/EC on waste (Marsden and Koivurova, 2011), have established a strategy for the effective management of C&D waste that includes the recycling and reuse of materials; currently, the estimated level of materials recovery (reuse or recycling) is less than 25% (Marsden and Koivurova, 2011). The next section will deal with waste management strategies based on recovery of C&D material in a more detailed way.

10.7 C&D reduction, reuse and recycling

Summarising the last section, there are two key ways of decreasing C&D waste: minimising the effect on the environment, including the health and safety of the operatives involved, and reducing costs, considering not only the using up of of landfill but also the potential of recovery (reuse or recycling). As a general remark, waste costs money (Johnston and Mincks, 1995). The following steps can be considered for determining if the recovery of materials is profitable (Peng et al., 1997):

1. recognise the potential materials
2. establish the costs and savings of recovery
3. propose a waste management plan
4. inform and train all the parties involved
5. implement the plan
6. monitor the plan
7. evaluate its results.

Taking into consideration the resource consumption and the potential environmental damage, there are several options that could be used, depending on the circumstances of the site, parties, contract, regulations and even the country. These options are, from the least potential environmental impact to the most (Peng et al., 1997): reduce, reuse, recycle, compost, incinerate and landfill. The first three are the most commonly used during the construction process, and they are generally known as the three Rs (Tam and Tam, 2006).

Reduction is the most efficient technique for reducing waste and its consequent problems; generally hazardous materials are the first target for this reduction because of the later increasing costs for waste disposal. This approach replicates the first strategy established by McDonald and Simthers (1998), employing techniques during the design, procurement and construction stages of the process. This entails a detailed understanding of the causes of R&D waste. Reuse moves materials from one site to another; in order to pursue this goal, a selective demolition technique moving down from the top of a building (for example) has to be applied mainly by hand or using time-consuming techniques (Poon et al., 2001). Recycling involves the reprocessing of the materials into new products that can generate additional benefits (Peng et al., 1997). C&D waste recycling is only feasible when the recycled product is competitive with the original one in relation to economy and capacity; generally the recycled material is an aggregate for lower-grade applications (Tam and Tam, 2006). The generation of 'new' raw materials through using or recycling materials decreases waste and, therefore, minimises environmental impact. Those materials that cannot be treated somehow are easily managed if only for the reason that there is less waste, even though they have to be incinerated or dumped.

Recycled aggregate can come from C&D materials. When concrete is fabricated using recycled aggregates, it is generally named as recycled aggregate concrete (Rao et al., 2007). This could be produced from recycled precast elements discarded after testing, or dismantled piece by piece, or from demolished concrete structures. In the first case, the aggregate can be relatively clean, whereas in the latter two cases the aggregate will be contaminated with many kinds of material, inert or not (dust, adhesives, salts, bricks, tiles, timber, plastics, cardboard, paper, metals, etc.). Besides the cost of the operation, in developing economies there still exists a poor image associated with the recycling activity as well as a lack of confidence in the finished recycled product (Rao et al., 2007). In general, the recycled aggregate concrete can always be used in lower end applications of concrete (fillings with mass concrete), and with previous testing and monitoring, it can be used as a regular structural concrete with the addition of fly ash, condensed silica fume and so on.

Regarding asphalt pavements, an important fraction of waste materials (not only the ones coming from C&D) have the potential to be used in road construction projects, such as glass, steel slag, tyres and plastics (Huang et al, 2007); reclaimed asphalt pavement materials can also be recycled for the same purpose. Current practice concentrates on the application of waste materials in the lower courses of the road (base and sub-base); they absorb materials in larger quantities than the upper courses, and regulations are not as restrictive. Nevertheless, according to Chiu et al. (2008), the most popular practices are the simple recycling of hot mix asphalt, the blending of ground tire rubber into asphalt (or asphalt rubber), and hot mix asphalt incorporating 10–25% of crushed glass for rehabilitating pavements (or glassphalt).

10.8   Environmental monitoring plan

The corrective measures that were proposed as a result of the environmental impact assessment process in the design phase should be implemented during the construction phase (Chen et al., 2005). It is also necessary to minimise residual impacts by making sure that the proper measures are implemented. In order to achieve this, current European legislation enforces an environmental monitoring plan (Marsden and Koivurova, 2011). Its main purpose is to establish a system that guarantees compliance with the corrective and protective measures and directions contained in the environmental impact statement. The establishment of an environmental monitoring plan involves monitoring the implementation of the proposed corrective measures, and also identifying any impacts which were not predicted in previous phases.

The construction company must present an environmental monitoring plan based on the environmental impact statement; it has to adapt to the company's characteristics and its current environmental management system (Lundberg, 2011). This plan comes into effect after it is approved by the environmental manager. The main objectives of the environmental monitoring plan are described in Box 10.2.

Box 10.2   Objectives of the environmental monitoring plan

- establish a system to coordinate and control the construction project, carried out to comply with: environmental protection, correction measures contained in the environmental impact study and also the determining factors in the environmental impact statement
- ensure that the activities are carried out in line with the environmental measures contained in the design project and the authorised conditions
- control the execution of the proposed environmental integration measures, and make sure that they are adapted to the determining factors given in the environmental impact statement
- confirm the efficiency of the corrective and protective measures and, if they are ineffective, determine the causes and take the necessary measures to solve the problem
- verify the quality of the materials and the execution of the project units proposed to ensure the environmental integration of the facility.

The most effective procedure involves establishing quantifiable indicators which can track the measures to be implemented. Normally the environment (pollution and noise), water and land are taken as reference indicators, although animal and plant species can also be used if the circumstances call for it. They should specify the location of the measurement points, the sampling method, their frequency and the parameters to be analysed. Comparison of the results obtained against the fixed standard allows deviations to be detected and reasoned conclusions to be drawn (Eccleston, 2011). All of this should be included in the periodic report (Ridgway, 1999).

10.9 Environmental impacts at the construction site

The objective of the corrective measures is to eliminate the effects of impacts, or at least minimise them as much as possible. It is possible to differentiate between the corrective measures implemented in the construction phase and the operation phase. It should be kept in mind that the design aims to harmonise functional, aesthetic and integration requirements. The efficiency of these corrective measures depends largely on their being applied at the right moment in time. The main measures to be taken during the execution of the construction project are described in Box 10.3 (adapted from Eccleston, 2011, and Jain, 2012).

> **Box 10.3   Measures to minimise environmental impacts during the construction phase**
>
> - keep the site as clean as possible, and protect certain areas with plastic or other types of material if necessary
> - ensure that areas where dust is liable to be produced are watered regularly
> - clean the parts of the site where mud may accumulate to prevent it spreading to the rest of the site
> - keep the works machinery in perfect condition to prevent excessive gas and noise emissions into the atmosphere
> - adequately design drainage on esplanades, roads and infrastructure to reduce land erosion
> - avoid blocking potential watercourses or surface runoffs
> - collect samples and establish an edaphic profile for later use
> - repair roads and esplanades which were dug up during the construction project
> - treat waste water before it is released back into the environment
> - replant vegetation in barren areas affected by the construction works, using native species, to avoid erosion and to integrate the works into the surrounding environment

(Continued)

- locate the facilities and machinery plant in places where they have a minimal visual impact, and take up as little space as possible
- restore and integrate pits, esplanades and any other areas which were used during the construction project
- remove all debris from the construction works in order to leave the site completely clean
- ensure that correct signalling is erected in all areas affected by the project
- avoid the passage of machinery and heavy vehicles through urban communities, particularly during maximum traffic times.

In general, the aim is to integrate the facility into its surroundings by means of environmental restoration. At times, it may be necessary to install acoustic barriers on roads and railways to prevent excessive noise in nearby urban communities. In other situations special measures must be taken to protect plant and animal species or even whole areas (woods, wetlands, natural parks, etc.). Other issues that must be taken into account beforehand are the effects on the historical-artistic heritage (whether visible or not): archaeological and paleontological remains, protected historic buildings, cultural heritage, etc. (Eccleston, 2011).

Focusing on one earlier point, cleaning of the area and waste collections are fundamental during the construction phase. It is often said that a clean site is a safe site. Cleanliness is also a reflection of the image of the main contractor. Many external observers may think that the level of cleanliness is an indication of how the company manages their business. However, it is not an easy task. The use of special cleaning and waste collection teams may be a good solution. It also provides a good opportunity to recycle materials: concrete, blocks, wood, metals and plastics. It is a difficult challenge for the construction site manager, which should be carried out successfully (Jain, 2012).

10.10 Sustainability in construction

Since the beginning of the 1990s, the main public and private organisations have believed that the construction industry should apply sustainability policies at all levels and phases. For the past 15 years, sustainable construction has been a common concept in any decision process within the sector, despite the fact that its practical applications have not been overly successful. The term 'sustainability' and also that of 'sustainable construction' have multiple definitions, each of them with different nuances (Hill and Bowen, 1997). It is generally considered that sustainable construction results in environmentally healthy facilities by applying ecological principles and using resources efficiently. Sustainability, when applied to construction industry, describes a process which begins in the feasibility and design phases and concludes in the facility's operation phase. It opts for the life-cycle focus, taking into account the interconnection between all the phases (feasibility, design, construction, operation and dismantlement). All of the above result in a demand for flexible facilities, which can be dismantled and which include recyclable components and materials. This life-cycle focus requires greater coordination by all the parties involved in the different phases of the construction process (Ridgway, 1999).

Some practical consequences of this new focus are (Van Bueren and De Jong, 2007):

- the enactment of new legislation to promote coordination throughout the facility's life-cycle
- the tendency to improve existing buildings in the surrounding areas instead of opting for new buildings
- the inclusion of the social system when analysing the surrounding areas
- the issuing of new technical codes based on performance instead of rigid procedures
- the introduction of new fiscal benefits
- the creation of sustainability certificates by certified bodies.

Hill and Bowen (1997) propose a series of principles, from different points of view (social, economic, biophysical and technical), the application of which may ensure that the construction sector is more sustainable.

10.11   Green buildings and certifications

The energy crisis of the 1970s led to the birth of a new stream of thought regarding the application of new designs and technologies for energy saving in buildings. Within this movement, the green building concept appeared. It is defined by the US Environmental Protection Agency (EPA, 2008) as 'the the practice of maximizing the efficiency with which buildings and their sites use resources –energy, water, and materials – while minimizing building impacts on human health and the environment, throughout the complete building life-cycle – from siting, design, and construction to operation, renovation, and reuse'. The strategy of this governmental agency regarding green buildings is focused on the following factors (EPA, 2008): energy efficiency and renewable energy, water efficiency, environmentally preferable building materials and specifications, waste reduction, toxics reduction, indoor air quality and smart growth and sustainable development. Smart growth seeks the involvement of the local residents in the development decisions in order to enhance their neighbourhoods, whereas sustainable development aims to reconcile economic growth and environmental protection.

There are several certification systems for ensuring the sustainability of a building. Currently, the two most commonly used in the construction industry are LEED and BREEAM (Rivera, 2009). They offer an independent third-party certification that assesses the performance of the facility against benchmarked key indicators (environmental, social and economic) in order to evaluate its specifications, design, construction and use. Each system recognises different levels of achievement. Both systems assess the performance of the building twice throughout the facility life-cycle: first at the end of the design phase, and then at the completion of the construction phase, but LEED requires both stages in order to proceed for certification, whereas BREEAM only requires one of them. Anyway, for each stage, the mandatory documentation has to be submitted to the certification body (USGBC for LEED and BRE for BREEAM) for review; if the evidence complies with the requirements, a rating is issued for the facility.

LEED (Leadership in Energy and Environmental Design) has been promoted by the US Green Building Council (USGBC) since 1995 (http://www.leed.net). This certification uses a suite of nine rating systems for verification of the design, construction and operation of green buildings and neighbourhoods. The rating indicators comprise five major categories: sustainable sites, water efficiency, energy and atmosphere, materials and resources and indoor environmental quality. Certification can qualify buildings in four levels: certified (40–49% compliance), silver (50–59%), gold (60–79%), and platinum (80% or more). Nevertheless, design is assessed against ASHRAE standards and guidelines, thus making it difficult to implement the LEED system outside the USA (Rivera, 2009). Some of the certified projects are the Bank of America Tower (New York City), the Clinton Presidential Library (Little Rock, Arkansas), the Queens Botanical Garden Visitor and Administration Center (Flushing, New York), or the Centre for Interactive Research on Sustainability (Vancouver, British Columbia).

BREEAM is promoted by the Building Research Establishment (BRE) and the UK Green Building Council (since 1990), as well as other European agencies in Germany, the Netherlands, Norway, Spain, Sweden and others (http://www.breeam.org). It stands for BRE Environmental Assessment Method. The assessed indicators include the energy and water use, the internal environment (health and well-being), pollution, transport, materials, waste, ecology and management processes. The levels of certification awarded are: pass, good, very good, excellent and outstanding. BREEAM is the environmental assessment method and rating system for buildings more used in Europe with more than 200 000 buildings certified; these include the new permanent venues for the 2012 London Olympic Games: the Olympic Stadium, Aquatics Centre, Velodrome, Media Centre, Eton Manor and the Handball Arena. BREEAM adapts well to every local or regional code and regulation. It is more flexible because of this fact, as well as the discretionary choice of being certified at just one of the stages previously mentioned, either design or construction (Rivera, 2009).

References

Aldana, J. and Serpell, A. (2012). Topics and tendencies of construction and demolition waste: a meta-analysis. *Revista de la Construcción*, 11(2), 4–16.

Chen, Z., Li, H. and Wong, C.T.C. (2005). EnvironalPlanning: analytic network process model for environmentally conscious construction planning. *Journal of Construction Engineering and Management*, 131(1), 92–101.

Chiu, C.T., Hsu, T.H. and Yang, W.F. (2008). Life-cycle assessment on using recycled materials for rehabilitating asphalt pavements. *Resources, Conservation and Recycling*, 52(3), 545–556.

Eccleston, H. (2011). *Environmental Impact Assessment: A Guide to Best Professional Practices*. CRC Press, Boca Raton (Florida).

EPA (2008). *Green Building Strategy*. United States Environmental Policy Agency, Washington DC.

Fatta, D., Papadopoulos, A., Avramikos, E., Sgourou, E., Moustakas, K., Kourmoussis, F., Mentzis, A. and Loizidou, M. (2003). Generation and management of construction and demolition waste in Greece: an existing challenge. *Resources, Conservation and Recycling*, 40(1), 81–91.

Formoso, C.T., Soibelman, L., De Cesare, C. and Isatto, E. (2002). Material waste in building industry: main causes and prevention. *Journal of Construction Engineering and Management*, 128(4), 316–325.

Gavilan, R. and Bernold, L. (1994). Source evaluation of solid waste in building construction. *Journal of Construction Engineering and Management*, 120(3), 536–552.

Hill, R.C. and Bowen, P.A. (1997). Sustainable construction: principles and a framework for attainment. *Construction Management and Economics*, 15(3), 223–239.

Huang, Y., Bird, R.N. and Heidrich, O. (2007). A review of the use of recycled solid waste materials in asphalt pavements. *Resources, Conservation and Recycling*, 52(1), 58–73.

ISO (2004). *Environmental Management Systems – Requirements with Guidance for Use (ISO 14001:2004)*. International Standardisation Organisation, Geneva.

Jain, R. (2012). *Handbook of Environmental Engineering Assessment: Strategy, Planning, and Management*. Elsevier, Oxford.

Johnston, H. and Mincks, W.R. (1995). Cost-effective waste minimisation for construction managers. *Cost Engineering*, 37(1), 31–40.

Kofoworola, O.F. and Gheewala, S.H. (2009). Estimation of construction waste generation and management in Thailand. *Waste Management*, 29, 713–738.

Lundberg, K. (2011). A systems thinking approach to environmental follow-up in a Swedish central public authority: hindrances and possibilities for learning from experience. *Environmental Management*, 48, 123–133.

Marsden, S. and Koivurova, T. (2011). *Transboundary Environmental Impact Assessment in the European Union: The Espoo Convention and its Kiev Protocol on Strategic Environmental Assessment*. Routledge, New York.

McDonald, B. and Simthers, M. (1998). Implementing a waste management plan during the construction phase of a project: a case study. *Construction Management & Economics*, 16(1), 71–78.

Peng, C.L., Scorpio, D.E. and Kibert, C.J. (1997). Strategies for successful construction and demolition waste recycling operations. *Construction Management and Economics*, 15(1), 49–58.

Pongrácz, E. and Pohjola, V.J. (2004). Redefining waste, the concept of ownership and the role of waste management. *Resources, Conservation and Recycling*, 40(2), 141–153.

Poon, C.S., Yu, A.T.W. and Ng, L.H. (2001). On-site sorting of construction and demolition waste in Hong Kong. *Resources, Conservation and Recycling*, 32(2), 157–172.

Rao, A., Jha, K.M. and Misra, S. (2007). Use of aggregates from recycled construction and demolition waste in concrete. *Resources, Conservation and Recycling*, 50(1), 71–81.

Ridgway, B. (1999). The project cycle and the role of EIA and EMS. *Journal of Environmental Assessment Policy and Management*, 1(4), 393–405.

Rivera, A. (2009). International applications of building certification methods: a comparison of BREEAM and LEED. *PLEA 2009–26th Conference on Passive and Low Energy Architecture*, 22–24 June, Quebec City, Canada.

Solis-Guzman, J., Marrero, M., Montes-Delgado, M.V. and Ramirez-de-Arellano, A. (2009). A Spanish model for quantification and management of construction waste. *Waste Management*, 29, 2542–2548.

Tam, V.W.Y. and Tam, C.M. (2006). A review on the viable technology for construction waste recycling. *Resources, Conservation and Recycling*, 47(3), 209–221.

Van Bueren, E. and De Jong, J. (2007). Establishing sustainability: policy successes and failures. *Building Research & Information*, 35(5), 543–556.

Further reading

Jain, R. (2012). *Handbook of Environmental Engineering Assessment: Strategy, Planning, and Management*. Elsevier, Oxford.

Morris, P. and Therivel, R. (eds.) (2009). *Methods of Environmental Impact Assessment (Natural and Built Environment)*. Routledge, New York.

11 Supply Chain Management

11.1 Educational outcomes

Outsourcing is very widespread in the construction sector. Adequate management of the outsourced resources by the main contractor allows to carry out the construction project more efficiently. By the end of this chapter, readers will be able to:

- comprehend the concept of supply chain management
- apply the concept of supply chain management to construction
- recognise the advantages and disadvantages of outsourcing in the construction industry
- understand the aspects related to procurement and management of construction subcontracts
- present the concept of 'lean construction'.

11.2 Introduction to supply chain management

Logistics aims to balance inventories between production capacity and customer service, taking into account trade-offs between the different functions of the process: purchasing, production, distribution and sales. In the 1980s, this traditional concept evolved into the supply chain approach when considering strategic management issues (Houlihan, 1988). The supply chain consists of a network of organisations that are involved in different processes and activities by way of multidirectional relationships, adding value as products and services for the final user or consumer (Vrijhoef and Koskela, 2000).

Construction Management, First Edition. Eugenio Pellicer, Víctor Yepes, José C. Teixeira, Helder P. Moura and Joaquín Catalá.
© 2014 John Wiley & Sons, Ltd. Published 2014 by John Wiley & Sons, Ltd.

Close cooperation develops between organisations in order to complement their different courses of action; this collaboration is based on beneficial association, being the result of the division of labour and specialisation (Kotzab et al., 2011). Therefore, supply chain management (SCM) is a term derived from the manufacturing industry, appearing contemporaneously with the concept of 'just-in-time' delivery. Cooper and Ellram (1993) defined SCM as 'an integrative philosophy for managing the total flow of a distribution channel from the supplier to the ultimate user'. The basic idea revolves around recognising interdependence in the supply chain and as a result improving its configuration and control, based on factors such as the integration of business processes. It involves the implementation of an integrated system for the management of operations and relationships in the supply chain.

Houlihan (1988) remarks on the differences between the SCM concept and the traditional logistic management, saying that SCM: 1. is viewed as a single process instead of several linked processes corresponding to the traditional functions; 2. depends on strategic decision-making; 3. views inventories as a last balancing mechanism; and 4. uses a systems approach, looking for integration instead of interfacing. Box 11.1 (summarised from Cooper and Ellram, 1993) shows further elaboration of these differentiating factors.

Box 11.1   Characteristics of supply chain management

- inventory management approach: aims for inventory reduction (especially if redundant) throughout the supply chain, but not necessarily as just-in-time systems
- cost efficiencies: there is an evaluation of costs along the supply chain in order to identify total costs advantages, thus allocating the savings to more productive uses
- time horizon: looks for durable (or indefinite) time relationships that turns members into loyal partners
- mutual information sharing and monitoring: members have to gain access to the information that they need in order to better manage their part of the supply chain
- coordination of multiple levels: there are three types of coordination to be considered (across supply chain members, management levels and functions), not just the typical single contact buyer–seller
- joint planning: many members in the supply chain must participate in the planning of material flows, new product development, etc.; over the years, the supply chain develops a continuous process of planning, evaluation and improvement
- compatibility of corporate philosophies: the members need to agree on the basic directions of the supply chain, but not necessarily in every step or procedure
- supplier base: this should be reduced in order to integrate more closely and to better coordinate the members of the supply chain

(Continued)

> - leadership: the supply chain needs leadership to develop and execute its strategy
> - sharing of risks and rewards: the members should be willing to share risks and rewards over the life of the supply chain
> - speed of operations: reduction of order cycle times increases the speed of operations; this can be achieved by using ICT throughout the supply chain, thus exchanging information on material flows and inventory, for example, among members.

11.3   The construction supply chain

Regarding the construction phase of the facility life-cycle, it is the main contractor who decides if internal resources are used to carry out different parts of the project, or alternatively, to acquire them externally from other more specialised companies. In the latter case, relationships are formed between the companies, which are regulated through subcontracts. Companies that work for the main contractor are generally called subcontractors or trade contractors; a specific term can be used to name the subcontractors of materials or equipment, referring to them simply as suppliers (Yik and Lai, 2008). In contrast with the subcontractors, the main contractor has been hired directly by the owner. However, the main contractor subsequently hires suppliers and subcontractors. The main contractor's role is to ensure that the contracts are fulfilled correctly.

At the same time, the subcontractors must make the same decision: use personal resources or hire a third party. In this way, a chain of firms can be formed, linked by successive contracts. So a number of tiers or layers of outsourcing are created: the firms that work for the main contractor (top-tier), and firms that work for these subcontractors (lower-tier); additional tiers of sub-subcontractors could exist, with self-employed workers being the final tier (Choudhry et al., 2012). This is known as the construction supply chain. In the construction industry, such practices are very common. For some works there may be various tiers of subcontracting: up to four or five tiers is common in Japan (Reeves, 2002) and some other European countries (Briscoe and Dainty, 2005). In Spain, a new law which came into effect in 2006 (Law 32/2006 on Subcontracting in Construction), allows only three tiers to be subcontracted, the self-employed workers being the final tier.

The construction supply chain can be differentiated at three levels (adapted from Yik and Lai 2008):

- primary level: only includes materials incorporated into the end product
- support level: provides equipment and machinery used in construction
- human level: provides labour.

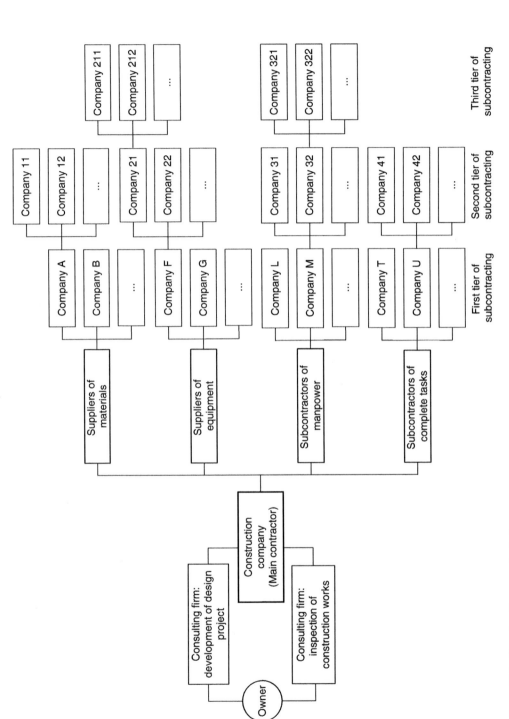

Figure 11.1 Construction supply chain.

There is a fourth possibility, which combines one of the first two levels (or both) with the provision of labour (Yik and Lai 2008). Figure 11.1 represents a typical construction supply chain; the owner and the consultants for designing and supervising the construction project are also included in the diagram. From this point of view, the 'upstream' supply chain (flowing from the owner) can be fundamental in fulfilling the objectives of the construction project. In this case, the main issue is the smooth flow of information between the parties involved: owner, consultants and main contractor (Choudhry et al., 2012).

Some characteristics relating to the construction supply chain should be also noted (Vrijhoef and Koskela, 2000):

- it is a convergent supply chain, given that all the supplies go towards the construction site where the facility is assembled, based on the supplies received. In the construction sector, the factory is based around a unique product and only one 'consumer' (the owner), in contrast with traditional production systems
- the supply chain is temporary, offering a unique facility by way of repeated organisational reshaping. As a result, the construction supply chain can be described as unstable and fragmented, and even more so due to the division between the design and construction phases
- it is a typical order-based supply chain, where every project generates a prototype. There is little repetition, although some of the processes may be very similar on some projects.

11.4   Pros and cons of subcontracting in the construction industry

Most of the companies in the construction industry contract out growing amounts of labour, machinery, equipment and material to third parties, particularly in the case of medium and large companies (Tam et al., 2011). Yik and Lai (2008) estimate that subcontracting represents around 50% of the gross production of the construction project, whereas for Vrijhoef and Koskela (2000) it accounts for more than 75%. Subcontracting is also a worldwide practice in the construction industry (González-Díaz et al., 2000; Tam et al., 2011; Choudhry et al., 2012).

From the point of view of the supply chain, substandard planning and last minute improvisations result in inefficient practices, which affect supply (Koskela, 1992 and 2003). Materials are ordered too late in some cases; the supplier is always in doubt as to the product demand and, as a result, applies buffers of safety to guarantee service levels. Sometimes, the materials are ordered too early, meaning that they are left in storage too long waiting to be used, giving rise to risks of deterioration and the problems inherent in storage and handling. Furthermore, the construction supply chain can be defined as non-collaborative, due to the adversarial relationships among the different stakeholders: owner, designer, main contractor and subcontractors: contractors and subcontractors do not look for cooperation but confrontation (Briscoe and Dainty, 2005).

From the point of view of the main contractor, the key motive for subcontracting is to transfer the risk to the subcontractor. The subcontractor can also transfer risk to the second-tier of subcontracting. This way, risks are transferred in cascade along the supply chain. This transference of risks might involve workload pressures, resources constraints and financial benefits (Beardsworth et al., 1988). The main contractor does not have to hire all the needed workforce, plant and equipment, thus keeping fewer full-time resources. Accordingly, the overhead costs are also reduced (Choudhry et al., 2012). Subcontracting provides specialisation and experience too; this division of labour increases the efficiency of every organisation involved at the construction site (Tam et al., 2011). Subcontracting can sometimes make it possible for the main contractor to start doing business in other regions using local subcontractors (González-Díaz et al., 2000). In summary, it could be acknowledged that subcontracting improves the flexibility of the whole supply chain (Yik and Lai, 2008).

A disadvantage of subcontracting from a business perspective is the loss of the underlying business benefits derived from that task (Beardsworth et al., 1988). Generally, subcontracted companies are smaller and have less financial power. The tasks they perform on a construction site require much more labour, and controlling the workers can be difficult. The working procedures are generally not well established in small construction companies, and this can occasionally give rise to fewer guarantees with regard to quality and professional competency (Tam et al., 2011). Additional tiers of subcontractors and suppliers make the communication along the supply chain more difficult (Yik and Lai, 2008). For this reason, the main contractor must supervise the work of its subcontractors more thoroughly, ensuring that they comply with the project's requirements (Elazouni and Metwally 2000); this fact can increase the overhead costs, thus reducing the benefits stated previously.

Vrijhoef and Koskela (2000) propose some practical initiatives to improve the supply chain in the construction industry:

- improve the interface between production activities on-site and within the supply chain, using techniques such as 'critical chain' or 'last planner' (to be explained in Chapter 12)
- improve the supply chain itself: the erratic and undisciplined nature of the owner's activities make it difficult to achieve this objective; there are problems at both ends of the chain, initially due to incomplete work definitions, and subsequently due to continuous changes in delivery dates
- transfer construction activities to the supply chain: prefabrication or industrialisation can be a partial solution, keeping in mind that it requires a more exact design process and in the case of error, correction is complicated. As a result, the prefabrication process is vulnerable to change

- integrate the construction project and the supply chain: the design–build contract is a simple example of this form of integration.

11.5   Procurement and management of subcontracts

Managing subcontracts is fundamental to every construction project. A large amount of the construction risk is transferred to the subcontractor and, as a result, contracts between the construction company and the subcontractors become very important (Beardsworth et al., 1988). Informal (or verbal) agreements are replaced by formal contracts, which include standard clauses binding the two parties (Civitello and Levy, 2005). Insufficient effort on the part of the subcontractor can affect the construction project objectives (timeframe, cost, quality, safety, etc.); therefore, subcontract management directly affects the project's performance (Tam et al., 2011).

Choosing a subcontractor requires the preparation of a tender process, which may vary in terms of complexity. The first step involves defining the scope of the work to be carried out, such that it can be adequately publicised to the interested companies. This definition must be as specific as possible to avoid subsequent problems during the execution of the works. It may be appropriate to attach shop drawings and technical specifications for the construction project, to eliminate gaps. The bid may be based on a fixed price for all the work contracted or be on a unit price basis, in line with the main contract. The awarding of the tender tends to be preceded by negotiations between the company and the different subcontractors, either face to face or by telephone. The subcontract is normally awarded to that company that bids a lower price while guaranteeing optimal quality; later, the contract must be prepared and signed (Halpin and Woodhead, 1998). Unethical practices should be avoided at all costs during the process, such as disclosing the price offered by one subcontractor to a rival subcontractor, to obtain a reduction. The process should be fair, to create an atmosphere of trust between the contractor and the subcontractor (Ballard and Howell, 2003).

The contract between the parties may include clauses adapted from the main contract (between the contractor and the owner) and other specifications which reflect the relationship between the main company and the subcontractor. The regular clauses generally included in the contract are shown in Box 11.2 (adapted from Civitello and Levy, 2005).

In any case, it is essential that clear working procedures be established between the company and the subcontractor, specifying conditions relating to time, cost and quality. The subcontracting procedure includes the following aspects (adapted from Halpin and Woodhead, 1998, and Civitello and Levy, 2005):

1. Establishment of a supplier database including basic contact data of the firm, documentary proof that the subcontractor complies with all

---

Box 11.2   Most common clauses included in the contract

- parties involved
- contractual documents
- scope of the work
- price
- payment terms
- changes, complaints and delays
- insurance and guarantees
- timeframes
- materials and labour
- party obligations
- disputes
- contract finalisation.

---

requirements demanded by regulations (external or internal to the company), accredited experience, evaluations of previous work if possible, or indirect references from other companies, clients, employees, etc.

2. Preselection of subcontractors based on the evaluations compiled in the database.
3. Specific definition of the service to be outsourced, by indicating general conditions and technical specifications relating to the construction project, as well as delivery and invoicing (see Box 11.2).
4. Requests and evaluations of bids.
5. Awarding the tender to the optimal bidder, according to the analysis carried out by the contractor.
6. Formalisation and signing of the order or contract.
7. Provision of permits and access cards to the construction site for the subcontractor.
8. Monitoring and control of the outsourced service (procedures and paperwork).
9. Final evaluation of the work and update of the database.

Box 11.3 details an example of procurement and management of subcontracts. Due to the requirements of the Spanish regulations, the first step for any subcontractor is to be approved by the main contractor. In order to get this approval, the subcontractor has to deliver documentary proof of its technical and financial solvency. Once this important issue is checked out, the steps explained in the previous paragraph follow. This process can be implemented in the contractor's organisation using an enterprise resources planning tool that allows the subcontractors to link to the extranet of the company to input data or check information (Griffith and Watson, 2004).

Box 11.3   A case study of implementation of a commercial ERP

A medium-size Spanish general contractor with an annual turnover of €500 million and 450 employees detects an important issue regarding subcontracting procurement and monitoring: the lack of internal procedure and control. There is a rigid normative framework in Spain regarding subcontracting in construction that details responsibilities and obligations of all the parties involved: there is even the specific Law 32/2006 on Subcontracting in Construction as well as Decree 1109/2007 that details it. Thus, the upper management asks for not only the detailed process but also for it to be implemented in the ERP of the company.

The process starts with the procurement of subcontractors (or suppliers if that is the case) for a particular activity of the project. There is a key step in the process at the procurement stage: only pre-approved firms can participate. In order to get a pre-approval, the firm has to provide documentary proof of its technical and financial solvency according to the Spanish regulations; this mandatory test limits and controls the liabilities between the parties. After all of the companies involved in the bidding have been checked out, additional steps follow: the awarding decision, the signing of the contract, the provision of permits and access cards to the subcontractor, the monitoring and control of any other procedure and paperwork.

This process was successfully implemented in the company using the commercial ERP. Furthermore, subcontractors were able to link to the extranet of the company in order to input new data or check information. In three months' time, the company was able to work with 80% of pre-approved firms; after one year of transition time, every subcontractor and supplier working with the company was already pre-approved. Administrative staff previously needed for the process were no longer needed, and were either transferred to other functions or made redundant. In two years, disputes and claims regarding the process of supply chain were reduced by 30%, and occupational accidents also diminished by 20%; however, there might have been other contributory causes of the decrease of these two ratios.

11.6   Purchase of materials and equipment

The purchase of materials is an area of subcontracting which does not include labour or specific equipment, and therefore performance does not depend on time. These materials are installed and set up directly by the main contractor, or alternatively by a subcontractor. As detailed above, it is also possible that the subcontractor may be directly responsible for the acquisition of materials

Code	Quantity	Material	Supplier	Delivery Date			Observations
				Proposed	Revised	Actual	

Figure 11.2   Material monitoring form.

(Yik and Lai, 2008). If the purchase depends directly on the main contractor, the transaction is called supply and the vendor is called the supplier. Three kinds of materials can be differentiated (Halpin and Woodhead, 1998):

- bulk materials that require simple or no fabrication and can be delivered from storage locations to the site at short notice – one to five days (fill materials, masonry, concrete, paving materials, water and waste piping, etc.)
- standard items that require some fabrication by the manufacturer, and they are stocked in small quantities, being manufactured for the project after the order is signed (e.g. bricks, ceramic tiles, standard steel elements, formwork, insulation products, valves, etc.)
- materials that are customised for a specific project and they must conform to unique requirements.

Specific contracts, called purchase orders, are normally used to formalise the acquisition, varying its complexity from a formal contract to a mail (email or fax) order (Halpin and Woodhead, 1998). In the case of complex and specifically fabricated materials or elements, the purchase order has to include detailed specifications (shop drawings, product data and samples) coming from the design project or developed from it by the main contractor or the project manager. Five basic items should be included in any purchase order (Halpin and Woodhead, 1998): quantity; description (as detailed as necessary); unit price; special instructions related (mainly) to delivery and payment of items (shipping and invoicing); and signature of the parties involved in the contract. Furthermore, the issues discussed in the previous section are also applicable to materials.

The order must also be adequately monitored. Confirmation that the items are delivered in the agreed condition is needed (see the proposed form in Figure 11.2). When the purchase is made from the construction site, the construction site manager is responsible for monitoring the order and confirming its reception (e.g. through a receipt). But many construction companies prefer to control and integrate the purchase of materials from the central offices (Griffith and Watson, 2004). They believe that, in this way, the company has greater control and a wider range of options to choose from, and as a result they can obtain a better price.

11.7   Coordination of suppliers and subcontractors

Many suppliers also subcontract part of the work to third parties. As detailed above, this practice is quite common, and four or five subcontracting tiers may exist in some cases. Initially, there is no contractual relationship between the main contractor and the second tier subcontractors. In addition, the main contractor also loses control over the work they do, given that the orders are transmitted through the first tier subcontractor. In any case this subcontractor is responsible for the work on behalf of the main contractor. The relationship between the higher and lower tier with contractors is similar to that which exists between the main contractor and the first tier subcontractor, so most of the concepts included in this chapter can be applied to this type of relationship (Choudhry et al., 2012).

It is very important that the main contractor's on-site personnel have adequate understanding of the work to be carried out by each subcontractor; this also applies to each subcontractor and its subcontractors on a lower tier. The nature of the work should be clearly explained to them, as should the contractual terms, the resources to be used and the safety measures corresponding to each subcontractor (Yik and Lai, 2008). Good coordination between both companies' personnel ensures that they can attain their objectives more easily, including their financial profit.

Periodic management meetings (e.g. weekly) between the main contractor and the subcontractors are necessary for carrying out the construction project correctly (Yik and Lai, 2008). The main objective of the meeting is the exchange of information regarding timeframes, overlaps between subcontracts, and the updating of information and documentation, including possible changes. A date should be set for the meeting and an agenda decided. The tools, techniques and procedures explained in Chapter 4 can be very useful in this context, for the main contractor as well as for the subcontractor.

Furthermore, the application of ICT throughout the construction supply chain is a current fact (Box 11.3). Tserng et al. (2005) explain the use of web-based technology, personal digital assistants, barcode scanning, and data entry mechanisms in a very detailed way, using a construction real-case. Once the main contractor has awarded each job to a particular supplier, it is decided which component or element will be traced or monitored using barcodes. There are six key phases where each element is monitored and controlled: production, test and storage, delivery, on-site inspection, inventory, and installation. Mobile devices scan the barcodes at the proper places and times, and the information is centralised through a web-based portal system. Data is thus acquired on-site, enabling monitoring of the operations on-line and control of the construction progress in real-time. There is also a feedback loop between the supplier and the contractor about the status of the construction project. Thus, rescheduling can be done if necessary, in a collaborative way. Information sharing is faster and more efficient; hence, issues can be detected sooner, misunderstandings can be avoided and corrective actions can be implemented faster in coordination between the parties involved.

11.8 Lean construction

During the 1980s, the Japanese automotive industry was booming, and it was more productive than the US and European industries and its cars were sold cheaper and in larger quantities worldwide. This fact drew the attention of a research group at the Massachusetts Institute of Technology. They started to compare and analyse practices used by Japanese manufacturers, especially Toyota and Nissan, and finally their research project summary was published in the book *The Machine that Changed the World* (Womack et al., 1990). In their final report, these authors revealed to western countries the Toyota Production System (TPS). This system is oriented to customer demand in order to develop services and high-quality products as efficiently and economically as possible, identifying and eliminating 'waste'. The TPS system is also strongly influenced by the work of Henry Ford (assembly line for the Ford Model T using standardisation and interchangeability of parts) and Edwards Deming (14 Principles of Quality as well as the Total Quality Management concept).

Later work of Womack and Jones (1996) and Liker (2004) gave a complete picture of the system. The TPS includes a philosophy of management, production and logistics, interaction with suppliers and customers, and a set of specific procedures and techniques. It has now been adapted and used by every car manufacturer in the world. Toyota never used the term 'lean' for the TPS system, although they did use the term 'just-in-time', but TPS is currently recognised worldwide as lean production, lean manufacturing or even lean thinking (Womack and Jones, 1996). The five goals of the lean production system are stated in Box 11.4 (Womack and Jones, 1996).

The construction industry is also characterised as being very traditional and conservative in reacting to change (Koskela, 1992). This is not only due to

---

Box 11.4   Five goals of the lean production system

1. Set the value desired by the customer for each product or service.
2. Identify those activities that add utility (value) to a product over all processes (value stream).
3. Make the production (and therefore also the value) flow seamlessly and continuously.
4. Implement pull production according to customer demand avoiding overproduction and waste.
5. Accelerate the improvement cycle, seeking perfection.

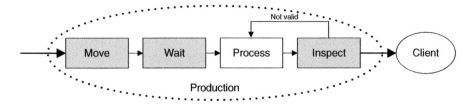

Figure 11.3   Production as a process flow (developed from Koskela, 2000).

the process, but also to its cultural context that largely influences the end user. A clear example of this is the use of prefabrication in buildings: in certain countries, consumers do not want prefabricated components, while in other countries they are commonly used. Nevertheless, it is important that innovative ideas, which aid efficient production and improve the product, are promoted as happens in the manufacturing industry.

Within this scenario in the construction industry, Koskela (1992) realised that the lean production system could be applied to construction. Production can be seen as a set of processes; the individuality of each of these processes is derived from a different design and control perspective. Koskela (2000 and 2003) believes that production is the flow of materials and information from the raw material to the end product. As part of this flow, the material can be processed, inspected and moved, or may also remain in waiting (Figure 11.3). The processing activities add value, while the others do not. As a result, lean construction is based on the flow of information and materials. These combine three different points of view on production (Koskela, 1992 and 2000):

- traditional: production is the conversion of inputs into products
- just-in-time: production is a logistical flow
- quality: production is the generation of value through satisfying the client's needs.

The objectives of lean construction are to reduce costs, reduce timeframes and increase value added for the developer. A broader definition of construction waste is used in lean construction: those resources that do not add value to the construction process or to the built facility (Formoso et al., 2002). Box 11.5 shows the ten basic principles for the improvement of construction processes applying the lean production principles (Koskela, 1992).

Conventional production is the transformation of inputs (materials, equipment, machinery, labour, etc.) into outputs (or products). Production can be divided into conversion sub-processes. From an accounting point of view, it is believed that minimum cost is achieved by minimising the costs in each sub-process. Conventional production, on the other hand, is improved by applying new technologies, mainly in those activities that present added value; activities which do not add value result in growing costs. In a lean construction framework, activities which do not generate added value are

> **Box 11.5   Basic principles for the improvement of construction processes**
>
> - reduce activities that do not add value
> - increase the product's value by considering the owner's needs
> - diminish variability
> - decrease cycle time
> - simplify and minimise stages
> - increase output flexibility
> - enhance process transparency
> - focus process control on the full production cycle
> - implement continuous process improvement
> - refer to processes ('benchmarking').

identified and a cost reduction process is initiated. On the other hand, activities that add value are improved in order to allow technological investments, which would develop them even further (Koskela, 2003).

According to Howell (1999), the lean construction approach differentiates from the traditional approach because:

- it establishes a number of clear objectives for the commissioning of the facility
- it looks for cooperation among all the interested parties
- it is oriented to maximise performance at the project level (in the general sense of facility life-cycle)
- it is applicable to production control throughout the facility life-cycle
- it is seeks for a concurrent design and execution of the facility.

Lean construction is a term coined in 1993 by the International Group for Lean Construction (http://www.iglc.net). It is in widespread use in some American countries such as Brazil, the USA, Chile, Peru and Colombia. To a lesser extent, modest actions have been taken in Europe: Finland, the UK, Germany, Spain and Portugal. Two applications of the lean construction philosophy are well known and commonly used:

- at the construction phase only (work-site): Last Planner System (Ballard and Howell, 1998 and 2003). It is a planning and control technique using different breakdowns whose main purpose, in addition to planning and controlling the production, is the reduction of work variability; three basic principles are applied: 1. Coordination of those who actually decide on site (last planners) through regular meetings; 2. Personal commitment to the final decision-makers; and 3. Public visibility of the results. This method will be detailed in Chapter 12

- at the life-cycle level (design and construction phases, basically): Lean Project Delivery System (also known as Integrated Project Delivery), a type of relational contract already explained in detail in Chapter 3 (Ballard and Howell, 2003).

There is much discussion regarding the application of lean production concepts in construction, with a large amount of scepticism in the sector with regard to its optimal implementation. It is argued that these concepts are applicable to repetitive processes but not to production by projects where each product and organisation is unique (Winch, 2006). Nevertheless, if every process is considered independently, repetitions may arise. Furthermore, the construction culture, in most of the scenarios and countries, is opposed to the implementation of innovative ideas. The application of this new management philosophy to construction requires drastic change, given that improvement efforts focused on costs, timeframes and productivity are not enough. Waste must be reduced and improvement principles applied at a global level: not only just for the construction works, but also for all processes, and this includes design, construction and operation.

References

Ballard, G. and Howell, G.A. (1998). Shielding production: essential step in production control. *Journal of Construction Management and Engineering*, 124(1), 11–17.

Ballard, G. and Howell, G.A. (2003). Lean project management. *Building Research & Information*, 31(2), 119–133.

Beardsworth, A.D., Keil, E.T., Bresnen, M. and Bryman, A. (1988). Management, transience and subcontracting: the case of the construction site. *Journal of Management Studies*, 25(6), 603–625.

Briscoe, G. and Dainty, A. (2005). Construction supply chain integration: an elusive goal? *Supply Chain Management: An International Journal*, 10(4), 319–326.

Choudhry, R., Hinze, J., Arshad, M. and Gabriel, H. (2012). Subcontracting practices in the construction industry of Pakistan. *Journal of Construction Engineering and Management*, 138(12), 1353–1359.

Civitello, A. and Levy, S. (2005). *Construction Operations Manual of Policies and Procedures*. McGraw-Hill, New York.

Cooper, M.C. and Ellram, L.M. (1993). Characteristics of supply chain management and the implications for purchasing and logistics strategies. *The International Journal of Logistics Management*, 4(2), 13–124.

Elazouni, A. and Metwally, F. (2000). D-Sub: decision support system for subcontracting construction works. *Journal of Construction Engineering and Management*, 126(3), 191–200.

Formoso, C.T., Soibelman, L., De Cesare, C. and Isatto, E. (2002). Material waste in building industry: main causes and prevention. *Journal of Construction Engineering and Management*, 128(4), 316–325.

González-Díaz, M., Arruñada, B. and Fernández, A. (2000). Causes of subcontracting: evidence from panel data on construction firms. *Journal of Economic Behavior & Organization*, 42, 167–187.

Griffith, A. and Watson, P. (2004). *Construction Management. Principles and Practices.* Palgrave MacMillan, London.

Halpin, D.W. and Woodhead, R.W. (1998). *Construction Management* (2nd Edition). Wiley, New York.

Houlihan, J.B. (1988). International supply chains: a new approach. *Management Decision*, 26(3), 13–19.

Howell, G. (1999). What is lean construction? *7th Annual Conference of the International Group for Lean Construction*, Berkeley, California.

Koskela, L. (1992). *Application of the New Production Philosophy to Construction.* CIFE Technical Report, 72. Stanford University, Stanford (California).

Koskela, L. (2000). *An Exploration towards a Production Theory and its Application to Construction.* Doctoral Thesis, Technical Research Centre of Finland, Espoo.

Koskela, L. (2003). Is structural change the primary solution to the problems of construction? *Building Research & Information*, 31(2), 85–96

Kotzab, H., Teller, C., Grant, D.B. and Sparks, L. (2011). Antecedents for the adoption and execution of supply chain management. *Supply Chain Management: An International Journal*, 16(4), 231–245.

Liker, J. (2004). *The Toyota Way: 14 Management Principles from the World's Greatest Manufacturer.* McGraw Hill, New York.

Reeves, K. (2002). Construction business systems in Japan: general contractors and subcontractors. *Building Research & Information*, 30(6), 413–424.

Tam, V.W.Y., Shen, L.Y. and Kong, J.S.Y. (2011). Impacts of multilayer chain subcontracting on project management performance. *International Journal of Project Management*, 29, 108–116.

Tserng, H.P., Dzeng, R.J., Lin, Y.C. and Lin, S.T. (2005). Mobile construction supply chain management using PDA and barcodes. *Computer-Aided Civil and Infrastructure Engineering*, 20, 242–264.

Vrijhoef, R. and Koskela, L. (2000). The four roles of supply chain management in construction. *European Journal of Purchasing & Supply Chain Management*, 6, 169–178.

Winch, G.M. (2006). Towards a theory of construction as production by projects. *Building Research & Information*, 34(2), 164–174.

Womack, J.P., Jones, D.T. and Roos, D. (1990). *The Machine that Changed the World.* Rawson Associates, New York.

Womack, J.P. and Jones, D.T. (1996). *Lean Thinking: Banish Waste and Create Wealth in your Corporation.* Simon & Schuster, New York.

Yik, F.W.H. and Lai, J.H.K. (2008). Multilayer subcontracting of specialist works in buildings in Hong Kong. *International Journal of Project Management*, 26, 399–407.

Further reading

Koskela, L. (1992). *Application of the New Production Philosophy to Construction.* CIFE Technical Report, 72. Stanford University, Stanford (California).

O'Brien, W.J., Formoso, C.T., Vrijhoef, R. and London, K.A. (2009). *Construction Supply Chain Management Handbook.* CRC Press, Boca Raton (Florida).

12 Resources Management

12.1 Educational outcomes

The management of resources on site is the primary function carried out by the construction site manager, as the party responsible on behalf of the construction company. The fact that this chapter is longer than the others in the book indicates its importance. After completing this chapter, readers will be able to:

- understand planning as a basic tool in managing construction resources
- present the work breakdown structure as a basic technique in planning
- define a project in terms of activities such that a schedule can be developed
- interpret techniques as timeline scheduling (or Gantt charts) and lines of balance
- describe the role and application of PERT/CPM for project scheduling
- develop a complete project schedule
- compute the critical path, the project completion time and its variance
- recognise time control as a counterpart to scheduling
- display the earned value method as a tool for planning and control, considering budget and schedule
- recognise the concepts of value engineering and risk management.

12.2 Construction planning

Turner (2008) adapts the five processes of management proposed by Fayol to the project approach: plan, organise, implement, conduct and control. In the case of the construction phase of the facility life-cycle, implementation corresponds to the execution of the construction works. Figure 12.1 shows the whole cycle (valid for each phase of the facility life-cycle).

Construction Management, First Edition. Eugenio Pellicer, Víctor Yepes, José C. Teixeira, Helder P. Moura and Joaquín Catalá.
© 2014 John Wiley & Sons, Ltd. Published 2014 by John Wiley & Sons, Ltd.

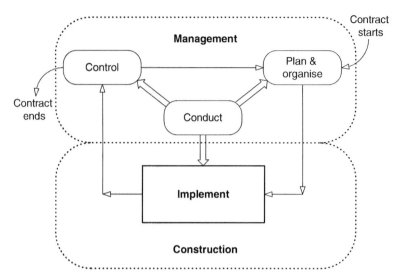

Figure 12.1 **The project management cycle (developed from Pellicer and Yepes, 2007).**

Box 12.1   Basic questions of construction planning

- what needs to be done? ⇨ Definition of project objectives, scope and requirements
- who will do it? ⇨ The project organisation: human resources (technicians for construction management and operatives for its execution) and subcontractors
- what additional resources are needed? ⇨ Materials, machinery and equipment
- when will it be finished? In what order? ⇨ Total and partial milestones
- how much will it cost (and when)? ⇨ Costs and budgets
- how will it be done? ⇨ Work procedures and control work performance

---

Construction planning consists of analysing all the activities to be carried out at the construction site within a particular time period, distributing and combining the available resources to optimise the cost, while maintaining the required quality levels (Edum-Fotwe and McCaffer, 2000). It involves the harmonisation of the time, cost and quality objectives with the available resources. Planning has to specify the project objectives, the strategy to achieve them (means and ends), and the standards to measure deviations. From the various alternatives available, the one that optimises the problem is chosen. Planning involves not only the traditional time focus, but also considerations of economics (costs and budgets) and methodology (work procedures). The basic questions to which planning should provide answers are provided in Box 12.1 (adapted from Nicholas and Steyn, 2008).

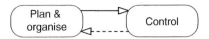

Figure 12.2   **The planning–control feedback loop simplified.**

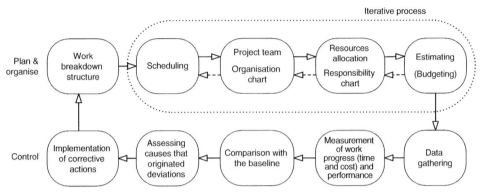

Figure 12.3   **The planning–control feedback loop in detail (developed from Pellicer and Yepes, 2007).**

Planning is the first step, but it is an insufficient process on its own to allow the construction project to be carried out within the proposed guidelines. There are at least three main reasons for planning the construction project: reducing uncertainty, improving efficiency and establishing a base for subsequent control. Through control it is possible to ensure that the construction works are in line with those planned and if there are any deviations, the necessary steps can be taken to put it back on course. The Project Management Institute (PMI, 2008) highlights three of this processes: plan, execution and control; considering only the managerial processes, the cycle can be simplified as displayed in Figure 12.2 that shows the planning–control feedback loop.

Planning is composed of several steps. The first is the definition of the project, which includes its objectives, scope, deliverables, specifications, applicable regulations (related to the kind of work, contract or country), user needs, constraints, assumptions and any other factor deemed important (Nicholas and Steyn, 2008). This is generally referred to as the statement of work and it is presented as a description of the work required for the construction project (Kerzner, 2009). Once this statement is set, the following actions have to be performed: 1. Determine the work breakdown structure; 2. Appoint the project team, according to an organisation chart, and establish the responsibility chart; 3. Schedule the project tasks; and 4. Estimate and budget the project tasks. The last three actions may have to be performed several times, iteratively, in order to meet the project goals (basically time, cost and quality). The main steps of the planning–control feedback loop are displayed in Figure 12.3; the steps regarding the control process will be explained in section 12.11.

Due to the characteristics of the construction industry, planning for the construction phase of the facility life-cycle has to consider some particular factors (Fisk and Reynolds, 2009):

- purchases orders in advance because of fabrication time, transportation time or material shortage
- interruptions and outages of utility services (water, power, gas, etc.), contacting the owner utility and determining the length, time, calendar dates, etc.
- temporary utility lines or temporary roads that need to be built to by-pass the construction site
- temporary utility service to the construction site: who provides it and where it is obtained from
- provision of contractor's work and storage areas, especially in urban environments
- traffic requirements and auxiliary works regarding movement of construction equipment, roads or streets closures, temporary access to the site and the like
- transportation and routes for special construction equipment
- coordination with other contractors working in the same area
- special regulations to be considered (environmental or labour).

12.3   Work breakdown structure

An activity, element or task is a definable piece of work, of limited duration, interrelated to but independent from others (the sum of the activities is the facility), measurable and to which a responsible person can be assigned (Kerzner, 2009). An activity requires human, material and financial resources if it is to be carried out. As the resources available within and outside the company are limited, the skills and experience of the construction site manager and his/her team determines what combination optimises compliance with a particular activity.

Usually the first step in the planning process involves the use of a technique called the work breakdown structure (WBS), which allows the work to be broken down successively into smaller elements of work (or simpler activities) aiming to manage the project adequately (Kerzner, 2009). The WBS consists of identification and hierarchical subdivision from the complete project into manageable tasks. The successive fractioning of the WBS is done through levels that give a greater degree of detail. The WBS is generally illustrated as a tree diagram, which reduces the complexity of the construction works by breaking them down into sets of activities. They can be broken down as much as necessary. Terminal elements or activities are the last level in the breakdown structure; that is the most detailed one. These terminal elements or activities are estimated based on the resources, time and budget; they are linked by interdependencies between them and can be scheduled.

The level of breakdown depends on factors such as type, size, resources, target schedule and control requirements demanded.

Each task is represented by a cell which contains all the necessary information for managerial purposes. A responsible party, optimal results, necessary resources, initial information and documentation, proposed time-frames and estimated budget may be assigned to each task. In the same way, the cell is identified using an alphanumerical code (Nicholas and Steyn, 2008). Box 12.2 includes a simplified example of the WBS for a construction project of an urban development.

Box 12.2   Simplified example of the WBS for a construction project of urban development

1. Layout
2. Demolition
3. Earth moving
    3.1   Clearing and grubbing
    3.2   Excavation and transport
    3.3   Embankments
4. Service facilities
    4.1   Excavation and transport
          4.1.1   Drainage and sewage
          4.1.2   Water
          4.1.3   Electricity
          4.1.4   Communications
          4.1.5   Gas
    4.2   Wiring and piping
          4.2.1   Drainage and sewage
          4.2.2   Water
          4.2.3   Electricity
          4.2.4   Communications
          4.2.5   Gas
    4.3   Inlets
          4.3.1   Drainage and sewage
          4.3.2   Water
          4.3.3   Electricity
          4.3.4   Communications
          4.3.5   Gas
    4.4   Landfills
          4.4.1   Drainage and sewage
          4.4.2   Water
          4.4.3   Electricity
          4.4.4   Communications
          4.4.5   Gas

(Continued)

5. Pavements
   5.1   Aggregate sub-base
   5.2   Concrete base
   5.3   Asphalt concrete
   5.4   Sidewalks
6. Gardening
   6.1   Sprinkler network
   6.2   Plantations
   6.3   Grass
   6.4   Urban furniture
7. Traffic signage
   7.1   Horizontal
   7.2   Vertical

12.4   Scheduling of activities

Scheduling is a part of the planning process that involves chronologically distributing the tasks over the available timeframe (Hinze, 2012). The schedule specifies the timeframe for the construction works in greater detail as it takes into account the time involved in various activities. Construction works scheduling therefore consists of predicting the resources to be used in the construction process and the point in time at which these various activities take place, such that they can be carried out, optimising costs, timeframe and quality. Once the activity schedule is designed, all useful information can be extracted from it, generally in the form of diagrams or tables: labour-time, investment-time and so on. The steps taken prior to scheduling of the construction project are displayed in Box 12.3. The following basic principles must be respected in activity scheduling:

1. Resources (labour and plant) must be levelled as much as possible (Winch, 2010). Underloaded and overloaded resources (either labour or plant) should be avoided; the first equals to waste, and the second produces premium payments, inefficient working because of burn-out effect or the use of marginal quality resources, and a high risk of delay in completing the tasks.
2. The resulting schedule should be clearly understandable, especially to those involved in carrying out the construction works; it is advisable that different parties use different approaches to the schedule (Mincks and Johnston, 2010).

Time is logically the basic variable in scheduling construction works. The schedule is used to deduce the optimal order for the execution of all the activities. It requires a series of basic steps (adapted from Kerzner, 2009): 1. Identify the interrelationships between tasks; 2. Establish the optimal time for the duration of

---

Box 12.3   Previous steps to schedule the project

- detailed study of the design project and the site where the construction works are to take place
- breakdown of the construction works into activities which depend not just on the type of work to be carried out, but also on the level of intensity assigned to scheduling, each activity measured and evaluated at current market prices
- analysis of the existing relationships between the different construction activities, based either on priority or dependency
- setting the construction methods and procedures to be followed
- for every activity, the construction procedure is studied, establishing the necessary equipment to execute them and calculating their performance levels.

---

these tasks; 3. Assign resources to these tasks and establish the resource histograms; 4. Obtain possible dates for each activity, gaps and the critical path; and 5. Verify and adjust the scheduling. The priority relationships express the interdependency between tasks that require earlier tasks to be concluded before these subsequent tasks begin. These links are necessary when it is not possible to execute two activities simultaneously due to physical restrictions, resource limitations, safety reasons, necessary waits or administrative processes (Goldratt, 1997).

To improve the efficiency of the planning process in construction projects, the schedule is generally developed through a hierarchy of three levels from low to high level of detail (Nicholas and Steyn, 2008). Ballard and Howell (1998) propose three levels: initial planning or master plan (long term), look-ahead planning (medium term) and commitment planning (short term). The long-term planning establishes the strategic goals and the main milestones of the construction project, defining few activities indexed at the top level of the WBS in order to comply with the established scope, time and budget. By using the medium-term planning, the schedulers optimise production according to the resources available at a lower level of the WBS, using more detailed activities. The short-term planning analyses and solves the conflicts and restrictions that may impede or hinder the fulfilment of the previously established objectives for the productive system; this commitment planning works at the lowest levels of the WBS, and consequently generates a great number of activities.

12.5   Duration of activities

The time available to carry out each activity or task is determined by the dates on which they are started and finished. The total time available corresponds to the hypothesis that all activities prior to the initial task are done as soon as possible, and all those subsequent to the final task are done with the greatest delay possible. The free time available is that allowed to carry out the activity, under the assumption that all the remaining, previous and subsequent activities are carried out as soon as possible. The independent time available relates to the idea that all tasks prior to the origin node are done as early as possible while all subsequent tasks are carried out as late as possible.

The concept of float is used to describe the amount of time that an event or activity can be delayed without causing a delay to subsequent tasks or project completion date. The float or work margin is obtained by subtracting it from the available time. In this way, total and free float can be defined, based on the corresponding available time. This information is very important for the project manager, to plan and control the project more actively and efficiently (Hinze, 2012).

A chain of activities linked by events, where each of them depends on the previous one, is used to define a path. The critical path is determined by all the events where the total float is zero. This determines the shortest time possible to complete a project. As a result, for the activities that comprise the critical path, the earliest and the latest dates coincide, both when starting and concluding each activity. The critical activities define the set of tasks that have to be carried out with no delay. All critical activities have a float of zero. Any modification in their duration results in an alteration to the critical path and, as a result, in the project's duration (Hinze, 2012).

The duration of an activity can be reduced by adding extra resources which, unfortunately, increase its cost. The resources assigned to each task can be modified to adjust them to the most appropriate conditions, based on the contingencies that arise during the project's execution. These changes result in an acceleration or deceleration in carrying out certain activities with the related increase or decrease in direct cost. As a result a relationship arises between the direct cost of each activity and the time invested in carrying it out, allowing the possibility of a cost–time adjustment, adaptable to the timeframe or financial investment needs at the time (Nicolas and Steyn, 2008).

The normal duration of a task is that which minimises its cost. At times, a schedule based on normal durations may excessively prolong the work, increasing the repercussions on the company's general project costs. In the same way, it is probable that the contractual term will be exceeded if scheduled solely with normal durations. In addition, there are reasons for shortening the duration of the activities of a project as much as possible ('crashing a project'): to free up resources to work on other projects, to exceed customer expectations, to make up for lost time and avoid contract penalties and to gain time-to-market competitive advantage, among others. In any case, the construction site manager may reduce the duration of some or all of the activities to decrease the total timeframe. Taking this to the limit, the crash point duration is obtained: it is the duration which minimises time while increasing resources. Every task requires a minimum execution time, even if infinite resources were available. It is expected that as the crash duration is approached, the incremental costs of reducing a task by one unit of time is proportionally

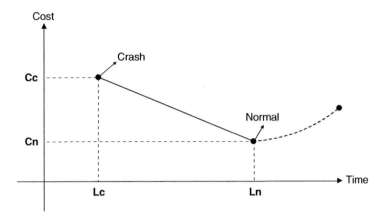

Figure 12.4 Time–cost curve for a task (developed from Nicholas and Steyn, 2008).

higher; thus, this curve, in fact, tends to have a hyperbolic shape, rather than linear (Nicholas and Steyn, 2008). To make this adjustment, the time–cost variation law must be known for each task, within some maximum and minimum duration limits. Figure 12.4 reflects the time–cost curve for a particular task, assuming a linear relationship.

12.6 Resources limitations and leveling

Every activity requires resources to carry it out.' In most real construction projects, scheduling without considering resource limitations may result in a non-credible schedule. Scheduling of available resources is a crucial issue for ensuring that the project is finished within the established timeframe and budget. It consists of associating resources with their respective tasks and seeing how they come together within the construction works. A graphic representation of the necessary resources over time is used (Figure 12.5); it is called a resource histogram. These histograms provide an efficient graphic means of observing trends over time and analysing low periods ahead of time (Nicholas and Steyn, 2008).

Resource limitations result in conflicts which could be solved either with situations of limited resources, which often leads to the extension of project duration, or the situation of unlimited resources within a time constraint. A levelling method evens out the resource histogram without making delays in the scheduled timeframe. However, allocation methods ensure that the necessary resources do not exceed those available, but on the condition that any resulting delay is kept to a minimum. With the help of various network techniques, a critical path is established and floats for each one of the activities are placed. (Nicholas and Steyn, 2008).

A variety of analytical and heuristic resource-constrained scheduling techniques have been developed to apply resource availability to the scheduling

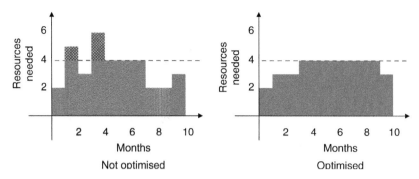

Figure 12.5   Resource smoothing with a fixed project duration.

process. Most project management software programmes employ heuristic methods because analytical methods require huge computational time, making them impractical for real construction projects. Heuristic methods provide good solutions in reasonable computing times, thus they are commonly used. The Burgess–Killebrew (1962) method for levelling and the Wiest–Levy (1977) method for resource allocation are examples of heuristic methods.

The Burgess–Killebrew algorithm (1962) is one of the pioneer algorithms in this field; it is also considered one of the most efficient. The resource histogram looks for non-critical activities that have the most advanced early completion date. This activity delays its completion by units of time until the float runs out. The earliest completion date for the activity is the one that minimises the sum of squares of resource usage. This is done with all the non-critical tasks, with priority given to those activities with greater float, where the most advanced early completion date for the two tasks coincides. Once done for every task, a new cycle of repetitions begins until such point as it is impossible to reduce the sum of squares any further.

The Wiest–Levy method (1977) is based on the scheduling of activities which can be carried out using the resources available. Nevertheless, the schedule can be revised in subsequent attempts. When the load is greater than the availability, an activity will have to be delayed; among the most non-critical activities, the one that solves the problem with the least delay is chosen. If there are two activities with the same conditions, the one with the greater float is delayed first, meaning that critical activities are delayed when there is no other option.

The following paragraphs provide a general description of common scheduling techniques: bar charts, networks and lines of balance.

12.7   Bar chart or Gantt diagram

Despite the fact that the Polish engineer and management researcher Karon Adamiecki developed in 1896 a tool called the 'harmonogram' (Packendorff, 1995), the well-known timeline bar chart was conceived by the American

TASK DESCRIPTION	1	2	3	4	5	6	7	8	9	10	11
A. Layout and facilities											
B. Demolitions			-----	-----	---- ●						
C. Earthmoving											
D. Service networks						-----	--- ●				
E. Pavements											
F. Signage											--- ●
G. Gardening											

Figure 12.6   Simplified example of a Gantt diagram.

engineer, Henry L. Gantt, between 1903 and 1917 (Wilson, 2003). This technique is quite elementary: it consists of a graphic representation based around two axes, the vertical axis featuring the tasks and the horizontal axis showing the time. Gantt attempted to solve the activity scheduling problem such that the duration of a basic task could be seen on the horizontal bar, including its start and completion date, and also the total time required in executing an activity. It is the most widespread scheduling method, as it adapts well to both small and large projects of all types, assuming that they are not overly complex. It is the most commonly used method in the construction industry also. It can be easily understood, even for those less familiar with any type of scheduling tools. The preparation of the bar chart requires a range of basic data, normally spread over columns, including:

- activities, in the order in which they are carried out
- budget or cost
- quantity in its corresponding units
- predicted performance for working equipment
- duration of the activity.

The timeline features on the horizontal axis. The time unit used may be days (short projects), weeks (medium-term projects) or months (long-term projects). The beginning and end of each horizontal bar represents the start and completion date for the corresponding task; the length of the bar is therefore proportional to the duration. The existing overlaps between the bars indicate the dates, from which the different activities begin simultaneously. The last two rows of the chart detail the cost or budget per unit of time in addition to that accumulated since the project's inception. Figure 12.6 includes a simplified example of the Gantt diagram for the WBS in Box 12.2, showing the tasks and their monthly distribution for the first breakdown level. Figure 12.6 shows the tasks which comprise the critical path (in dark colours) and the float computed using the critical path algorithm, as explained later.

Bar charts are very effective in the initial planning stages, but the graphic can become confusing when changes are made. For complex projects it has serious limitations. It was these difficulties that gave rise to the development of network diagrams.

12.8 Network diagrams

12.8.1 Historical introduction

Network diagrams identify the interrelationships between tasks and establish the most appropriate moment for their execution. They also help in preparing the project calendar and determining critical paths. The typical network represents a set of different 'arrow diagrams' which go from the origin node to the destination node. In this sense, the path is defined as a sequence of connected events which flow from the start of the project to the end. The time necessary to cover any of these paths is the sum of the time corresponding to each of the tasks involved. The critical path is the one that requires a longer period of time to progress from start to completion and indicates the minimum timeframe necessary to complete the whole project.

The critical path is essentially the route which represents a project's bottleneck. The reduction of the total execution timeframe will only be possible if the activities on this path can be shortened, since the time necessary to execute non-critical activities does not affect the project's total duration. The habit of accelerating all project activities in order to reduce the total timeframe is therefore unnecessary. The decrease of the execution term, for one or various critical activities, can reduce the project's total timeframe; it may also change the critical path such that activities which were not previously critical become so. There are many variants within this set of scheduling techniques. The classic Critical Path Method (CPM) has been widely used for project planning ever since its invention in the 1950s. This technique was developed by Morgan R. Walker, of the Engineering Services Division at Du Pont, and James E. Kelley, who worked at Remington Rand. Walker and Kelley were interested in solving the problem of improving scheduling techniques used in the construction of industrial plants; it is based on a deterministic or exact duration of each task (Kelley, 1961). Deterministic CPM is easy to understand and use, for the purpose of project control. A survey of the construction industry regarding the use of CPM scheduling for construction projects can be found in Galloway (2006).

The other technique, the project evaluation and review technique (PERT) was originated by the US Navy, when Admiral W.F. Raborn recognised that an integrated planning system and reliable control system were necessary for scheduling Polaris ballistic missiles. With his support, a research team was established in 1958 to develop PERT. At the time of the Navy's first internal report on the matter, PERT became the *programme* evaluation and review technique. D.G. Malcolm, J.H. Roseboom, C.E. Clark and W. Fazar, all from the research team sponsored by the Navy, were authors of the first document published on PERT (Malcolm et al., 1959). This method is based on the probability of activity durations and is best suited to reporting on works in which major uncertainties exist.

Both techniques (CPM and PERT) offer the necessary components to form the current critical path, using the control of implementation periods and

Activity	Depends immediately on	Duration (months)
A. (Layout and facilities)	–	1
B. (Demolition)	A	1
C. (Earthmoving)	A	4
D. (Service networks)	B	4
E. (Pavements)	C	4
F. (Traffic signage)	E	1
G. (Gardening)	E	2

Figure 12.7   List of activities and priority relationships.

operating costs, to execute a project in the least time and at the lowest possible cost. These techniques differ mainly on how the duration of activities is processed.

12.8.2   Graphical representation

The graphical representation of a project is called a network and is composed of a list of activities and priorities based on accepted principles. There are two types of common activity network models: the precedence diagramming method (PDM), also called activity-on-node network (Fondahl, 1961), and the arrow diagramming method (ADM), also called activity-on-arrow network (Kelley, 1961). These networks were developed as a reaction to the growing complexity of projects and introduced logic relating to the dependency of activities on each other. This visual representation allows each part to be related to the whole, allowing the project to be easily understood. It is important that the priority relationships between the activities involved in the project are clearly identified and represented using partial graphs, which will subsequently be used to form a complete network. To prepare the network, a table must be fill in as shown in Figure 12.7, based on the example in Box 12.2.

A network diagram includes the following steps (Griffis and Farr, 2000):

- all the project activities should be clearly defined and identified
- the technological sequence between the activities must be indicated
- a network must be constructed which shows the relative priority relationships
- the execution periods for each activity should be estimated
- the network is evaluated once the critical path is calculated
- as time goes on, and more experience is acquired, it is revised and evaluated.

The typical network characterises a set of different arrow diagrams which go from the origin node to the destination node. In this sense, the path is defined as a sequence of connected activities, which flow from the start of the project (node 1) to the end. The time required to follow one of these paths is the sum of the times corresponding to each of the activities. The critical path is that which requires the longest time to progress from inception to completion, and indicates the minimum timeframe required to complete the whole project.

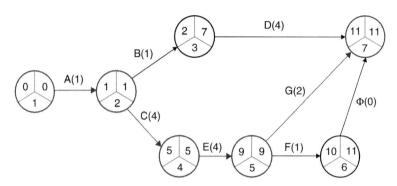

Figure 12.8   Arrow diagram for a project representing its critical path.

Figure 12.8 is a graphical representation using the arrow diagram technique of the example described in Box 12.2. Note that a dummy activity(Φ) of no duration had to be entered so that activities F and G did not have the same start and finish node; only in this way it is possible to differentiate the two activities and calculate the float for one of them, in particular that of F.

12.8.3   Calculating the critical path

In the classic CPM analysis, an algorithm is used to calculate the critical path, which involves first a forward pass calculation and then a backward pass calculation through the network (see Griffis and Farr, 2000). The forward step starts at the origin node and ends at the destination node. At each node a number is calculated, which represents the earliest start time for the corresponding action. These numbers are represented in Figure 12.8 in the upper left sector. For the backwards step the calculations are taken from the destination node and flow towards the origin node. The number calculated at each node (shown within the upper right sector) represents the latest finish time for the corresponding activity.

The early start time for each activity is the earliest date on which this activity may begin, assuming that the priority activities have started on their respective earliest start dates. As discussed, the forward step begins at the origin node, where D_{ij} is the duration of the activity (i,j) and E_i the early start time for all the activities which share origin node i; if $i=1$, therefore it is the origin node and by convention $E_1=0$ for all critical path calculations. The earliest possible occurrence time for an event is the largest of the early finish times for those activities which terminate at the event. Thus, the calculations for this forward step are obtained using the equation 12.1:

$$E_j = \max_i \left\{ E_i + D_{ij} \right\} \tag{12.1}$$

If the time necessary to carry out an activity is D_{ij}, the early finish time can similarly be determined as $T_{ij} = E_i + D_{ij}$.

The last finish L_j, is the latest an activity can finish without delaying the project beyond the deadline. In the same way, the last start I_{ij}, is the latest an activity can start without delaying the project's completion date; it is defined as $I_{ij} = L_j - D_{ij}$.

The second phase, the backwards step, begins at the destination node. Its objective is to calculate the last finish (L_i) for all activities which conclude at node i. If $i = n$, then this is destination node, $L_n = E_n$, and corresponds to the start of the backwards step. The latest permissible occurrence time for an event is the minimum (earliest) of the latest start times for those activities which originate al the event. Thus, for any node i:

$$L_i = \min_j \left\{ L_j - D_{ij} \right\}$$ (12.2)

Once the two phases are complete, the activities which comprise the critical path can be identified; they are those which satisfy the following conditions:

$$
\left.
\begin{array}{rl}
\text{(i)} & E_i = L_i \\
\text{(ii)} & E_j = L_j \\
\text{(iii)} & E_j - E_i = L_j - L_i = D_{ij}
\end{array}
\right\}
$$ (12.3)

But the float calculations provide interesting information on the project. There are two important types of floats (Griffis and Farr, 2000): total and free. The total float, also referred to as slack, H_{ij}, for activity (i,j), is the difference between the maximum time available to carry out the activity $(L_j - E_i)$ and its duration (D_{ij}); it represents the maximum time that the activity can be delayed without extending the completion time of the whole project:

$$H_{ij} = L_j - E_i - D_{ij} = I_{ij} - E_i = L_j - T_{ij}$$ (12.4)

In terms of the free float, it is assumed that all activities start as early as possible. In this case, the free float, F_{ij} for activity (i,j), is the excess of time available $(E_j - E_i)$ over its duration (D_{ij}); it represents the maximum time that an activity can be delayed without affecting the early start of any succeeding activity. With free float available for an activity, a project manager knows that the float can be used without changing the status of any non-critical activity to become critical. An activity which has a positive total float may also have free float, but this can never be greater than:

$$F_{ij} = E_j - E_i - D_{ij}$$ (12.5)

In the example represented in Figure 12.8, it can be seen that the activities that make up the critical path are A, C, E and G, turning out a project with an 11 month duration. Activities B, D and F have floats, both total and free, of 3, 2 and 1 month respectively.

12.8.4   Probability applications

Since the end of the 1950s network methods have been applied successfully to the scheduling of activities in industry and construction, increasing the probability of a project finishing on time (Galloway, 2006). Nevertheless, one of the inconveniences of CPM is its deterministic character. This focus is not appropriate for projects where it is difficult to estimate an exact duration for each activity.

Time in reality is a random variable which has certain degrees of probability. In this sense, the original version of PERT required the use of three different types of estimations on activity time to obtain basic information with regard to its probability distribution. This information is used for all activity times to estimate the probability of completing the project on the proposed date. In this case, uncertainty and risk are associated with scheduling through the use of minimum (or optimistic), maximum (or pessimistic) and hopeful (or likely) duration concepts for each task. The central limit theorem of statistics allows the estimation of the probability of fulfilling the project timeframe based on average values and variances in the activities which comprise the critical path. In this way, having determined the optimistic, pessimistic and likely durations for a task, it is possible to estimate its average duration and variances considering a type β distribution function (Griffis and Farr, 2000; Nicholas and Steyn, 2008). Assuming that the number of activities involved in the critical path is sufficiently high, that the duration distribution functions for each task are of the same type and are statistically independent, it is possible to estimate the project timeframe based on the sum of the activity durations which comprise the critical path. In addition, the variances in the project's duration are evaluated using the sum of the variances in activities which form the critical path, using the Gauss curve distribution function. These conditions are sometimes not fully followed on projects, but even so, the PERT technique has offered good practical results. In this sense, Ahuja et al. (1995) claimed that the PERT calculated mean project time is always an underestimate of the true project mean. Lu and AbouRizk (2000) proposed a simplified CPM/PERT simulation model by taking a discrete event modelling approach.

The three time estimations used by PERT for each activity, to provide most likely, optimistic and pessimistic estimations, are (Malcolm et al., 1959):

- optimistic time (b): best expected time if the activity's execution proceeds better than is normally expected
- pessimistic time (a): worst expected time if everything goes wrong (but excluding major catastrophes)
- most likely time (m): expected time, assuming everything proceeds as normal.

Due to the fact that 'a' (optimistic time) and 'b' (pessimistic time) may vary in relation to 'm' (most likely time), this distribution may be asymmetrical.

As the β distribution was the one that best adjusted to the general properties, it was chosen by the original PERT research team as a model to determine the average expected times t_e and the typical or standard distribution σ, associated with the three time estimations (Malcolm et al., 1959; Clark, 1962). Following an analysis which involved a hypothesis on the relationships between the route and typical deviations, and an approximation with regard to the relationship between the average and mode in the β distribution, the following general formulae were proposed for t_e and σ:

$$t_e = \frac{\frac{1}{2}(a+b)+2m}{3} = \frac{a+b+4m}{6} \tag{12.6}$$

$$\sigma^2 = \left(\frac{b-a}{6}\right)^2 \tag{12.7}$$

12.8.5 The precedence diagramming method

The activity-on-arrow (AOA) network has some disadvantages and limitations. This AOA network can only show finish-to-start relationships. Also, it is necessary to use dummy activities to show some of the more complex relationships between activities that indicate the dependency of one task on other tasks but for other than technical reasons. These limitations are important, for example, in linear projects such as highways, pipelines and canals, where activities must be split, increasing their total number, and making the schedule more complex to model. Because of the above limitation, Fondahl (1961), as part of a Stanford University team, developed an activity-on-node diagramming method; this method was later known as precedence diagramming. This network model is more flexible than AOA, includes realistic relationships between activities and allows overlaps (Hinze, 2012).

The precedence diagramming method (PDM) uses nodes to represent the activities, connecting them with arrows to show the relationships that exist between different activities. It includes four types of dependencies or precedence relationships (Hebert and Deckro, 2011):

- finish-to-start (FS): an activity cannot start before a previous activity has ended. This type of dependency is the one most commonly used
- finish-to-finish (FF): an activity must finish before the next activity can finish, i.e. both activities should finish simultaneously
- start-to-start (SS): an activity must start before the next activity can start, i.e. both activities should start simultaneously
- start-to-finish (SF): an activity must start before the next activity can finish. In the real world, this type of dependency occurs less frequently than the other ones.

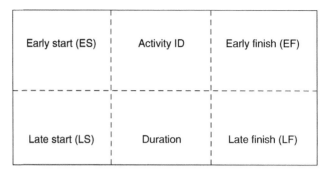

Figure 12.9   PDM activity node labels.

PDM represents the construction activity as a rectangle (Figure 12.9). In each box there are six items of information: the activity identifier, the activity time (or duration), the early start (ES), the early finish (EF), the late start (LS) and the late finish (LF). The activity identifier is a short alphanumeric label assigned to each schedule activity. The duration is the total number of work periods (in calendar units) required to accomplish the work of the activity. Based on preceding logic and constraints, ES is the earliest point in time when the schedule activity can begin, EF is the earliest point in time when the schedule activity can complete, LS is the latest point in time when the schedule activity can begin so as not to delay the project completion date or any constraint, and LF is the latest point in time when the schedule activity can finish, so as not to delay the project completion date or any constraint (Hinze, 2012). Each box is connected with another one according to the relationship (links) between those activities, including a lag where is necessary. The scheduling of the model is only effective as the model represents reality.

The time analysis for precedence involves both a forward pass and a backward pass, as in ADM. The forward pass evaluates all activities preceding a given activity to determine its ES and EF times and the minimum total project duration. The process assumes that all activities will start as soon as possible. Moreover, the activities will finish as soon as possible; subsequently, succeeding activities will start without delay. EF is found by simply adding the duration of the activity to its earliest start. If more than one activity precedes an activity, the ES comes from the LF of all preceding activities. Adding other relationships than finish-to-start and leads and lags significantly complicates the process, but the definitions of ES and EF remain the same (Hinze, 2012).

The objective of the backward pass is to calculate how late each of the activities can start and finish without delaying the completion of the overall project, or any imposed constraint. In this case, we start at the end of the schedule and logically work from the finish to the beginning evaluating the LF and LS time for all activities and milestones. In this case, we assume that all activities will start and finish as late as possible; each activity will finish as the earliest of its successor starts. If there is a single succeeding activity LS is found by simply subtracting the duration from the LF. In case of multiple

Activity	Duration (days)	Predecessors	Precedence relationship
A	5	B C D	FS+0 FS+0 SS+15
B	10	C E	FF+3 FS+3
C	5	D F	SS+5 FS+20
D	20	F	FS+2
E	30	G	FF+2
F	5	G	FS+7
G	10	—	—

Figure 12.10   Project network example for seven activities.

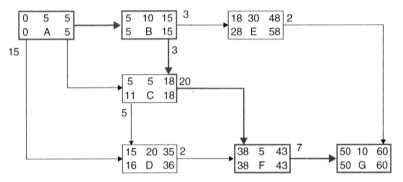

Figure 12.11   Scheduled precedence network (evaluated from Figure 12.8).

successors, the LF is the LS of all succeeding activities. Other relationships and leads and lags do not modify the definitions of LS and LF (Hinze, 2012).

To illustrate the precedence method of just seven activities Figures 12.10 and 12.11 show the duration and relationships of these activities and the corresponding scheduled precedence network.

12.8.6   Critical chain

Network techniques are commonly used for large projects. For these techniques the critical path is the ordered sequence of tasks that determines the longest duration of the project, which has the following characteristics (Wiest, 1964):

- there may be more than one critical path
- each of the tasks on the road is called 'critical' and has no buffer
- a delay in any of the critical tasks also delays the project completion
- other tasks can loosely be delayed to consume their float without causing a similar effect.

These techniques implicitly assume that there are unlimited resources available for allocation to project tasks. However, when resources are limited, the task can be delayed because of unavailability of resources or for technological reasons. In that case, a critical sequence of work on a project can be identified (Wiest, 1964). Unlike the critical path, a critical sequence is determined not only by technological issues or task priority, but also due to resource constraints. Just as on the critical path, the work is located in the critical sequence with zero clearance, and the length of the critical sequence determines the minimum duration of the project.

Moreover, Eliyahu Goldratt developed the theory of constraints (Goldratt and Cox, 1984). This theory states that any system has a restriction that limits its production. Once the main objective of the system is set, and it is accepted that there are restrictions on the system, the goal is to eliminate them. Constraints can be physical in nature (e.g. human, material, supplies, etc.) or not (e.g. procedures, morale, training, etc.). Applying the theory of constraints to the project led to the concept of critical chain (Goldratt, 1997), similar to that of critical sequence proposed by Wiest (1964).

Critical chain looks to detect the longest chain of dependent events, considering it a restriction. Normally, there is a resource intervening in multiple activities that will determine the schedule of the project. In order to protect the critical chain, resource buffers are established. This method also eliminates activity buffers, concentrating on a project buffer and introducing feeding buffers in the non-critical sequences. No complex algorithms are proposed but a focused management of uncertainty through the appropriate use of human resources management (Goldratt, 1997). An in-depth explanation of the method can be found in Leach (2000).

12.8.7 Commercial software

There is a lot of commercial software on the market for project scheduling, some simple and economical, others more complex and expensive, but generally they schedule by using both bar and network methods. Regarding network techniques, most of the commercial software available uses the precedence diagramming method, so this method is the most widely used currently by practitioners (Kerzner, 2009; Hinze, 2012). Graphically, the nodes represent the activities, and the arrows their interrelationships. These methods allow a more complete understanding of the bar chart for all the different tasks involved in a project. They provide information on the dependency relationships and the decisions to be made to reach the proposed aim. Fundamentally they provide the following information:

- tasks to be carried out to reach the objective in the proposed timeframe
- activities to determine the project timeframe (critical path)
- minimum project duration based on the resources available
- tasks which require a special effort to reach the established timeframe

- current project situation with regard to the completion date
- diagram of dynamic activities, which can be modified in line with the data relating to the construction works
- the effect of the changes made to a schedule on the construction works activities and phases which remain to be carried out.

The two most popular software tools are Primavera Project Planner™, by Oracle Co., and Microsoft Project™. According to a survey developed by Galloway (2006), Primavera was preferred by 65% of the respondents, whereas MS Project was chosen by 20%. However, the general opinion was also that Primavera is more complex and difficult to understand, thus increasing the cost of managing the project.

12.9   Line of balance

The 'line of balance' was conceived by Goodyear Co. at the beginning of the 1940s and developed further by the US Navy in the early 1950s (Arditi et al., 2001); it has been known by other names, such as velocity diagrams, linear scheduling method, time-space scheduling or repetitive project model (RPM). It is used fundamentally for tasks of a recurring or repetitive nature, based on the hypothesis that performance of a particular task is constant, and therefore this performance is linear in relation to time. Repeated tasks allow greater opportunities to increase productivity over the course of time and due to experience and practice. Learning is a phenomenon which occurs when the time and effort required to complete repetitive tasks are reduced as repetition increases (Arditi et al., 2001).

It was first implemented in industrial manufacturing, aiming to attain a constant flow rate of finished products. Nevertheless, repetitive activities can also be found in construction (Al Sarraj, 1990): high-rise buildings, housing projects, highways, pipeline networks, bridges, tunnels, railways, airport runways and water and sewer mains. In these construction projects some resources are required to repeat an activity throughout the project, moving from one place to another. Therefore, scheduling these activities must look for work continuity that allows the crew to finish the work in one place and move to the next, minimising idle time and maximising the use of the resources.

Their main application to construction occurs when one of the three dimensions dominates over the others. This may be height, in the case of buildings, or length in roads, railways, electrical lines and piping. Linear civil engineering infrastructures, mainly roads and railways, generally name it a time-space chart. This graphical representation reflects the longitudinal route of the linear works on the horizontal axis; on the vertical axis the chart shows, for example, the phases, the schematic plan, kilometres, characteristic profiles, milestones, structures, earthmoving work (longitudinal compensation or mass diagram – see Box 10.1 in Chapter 10). All of them are used to draw the activity graphically,

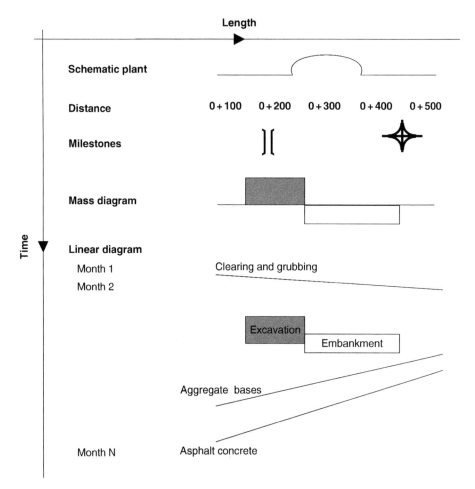

Figure 12.12   Simplified example of a time-space chart.

where every activity is given coordinates in space-time. The activities executed on a continuous linear basis are represented by way of a line (Harris et al., 2006). Other activities can be initiated and end at the same time throughout a piece of the construction site; they are shown as rectangles. Figure 12.12 represents a schematic example of a time-space chart for a road project.

12.10   Last planner system

The last planner system (LPS) is a scheduling and control technique developed under the lean construction philosophy (see section 11.8 in Chapter 11), probably most well-known and used than the lean construction philosophy as a whole. It is focused on the construction phase of the facility. It was proposed by Glenn Ballard and developed by members of the Lean Construction Institute

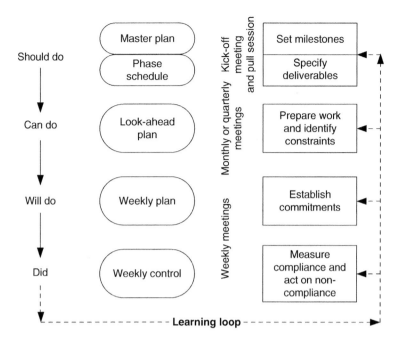

Figure 12.13   Scheme of the last planner system (Rodriguez et al., 2011) (Reproduced with permission from Colegio de Ingenieros de Caminos, Canales y Puertos).

in the USA (Ballard and Howell, 1998 and 2003). LPS is not a technique that aims to replace or compete with traditional methods of bars and networks, but it complements and enriches them decreasing variability and improving workflow. LPS is specially designed to increase the reliability of the planning process and, therefore, improve the performance of the project.

Different persons in several organisations and at different moments of the project are involved in the planning process. They are named last planners. LPS defines criteria for prospective production commitments to protect the activities of uncertainty and variability. The implementation process of the system is performed in cascade, using a hierarchy of three levels from low to high level of detail, as explained in section 12.4 and shown here in Figure 12.13 (Rodriguez et al., 2011, developed from Ballard and Howell, 1998 and 2003):

- review the master plan or initial planning (long term) that establishes the strategic goals and the main milestones of the project; in case of large or complex projects, develop the schedule of each phase
- develop a look-ahead planning (medium term) with a time horizon of one to three months that optimises production according to the available resources and then analyse constraints in order to eliminate bottlenecks, using more detailed activities framed within the master plan

- deepen the plan through commitment planning (short term) through weekly (or sometimes fortnightly) meetings of the last planners (site managers, foremen, subcontractors, key suppliers, etc.) who analyse and solve the conflicts and restrictions that may impede or hinder the milestones established at the previous levels, generating a large number of detailed activities
- verify the compliance with the weekly plan, identifying the causes of non-compliance.

Once the master plan is set, the first meeting of the last planners (site managers, foremen, subcontractors, key suppliers, etc.) is a collaborative session called a 'pull session'. The master plan is reviewed, starting from its final deadline, backwards. The logic of the plan is analysed through brainstorming of the last planners. Every one of them has to state the constraints of their job, preceding task that should be finished, estimated duration, resources and so on. Once there is a draft schedule, it has to be re-examined – in depth, if need be – to shorten the tasks in order to fit the key milestones of the master plan. The session has to finish by summarising the agreements reached by the team. Look-ahead sessions will be held among the last planners every month (bi-monthly or quarterly) to realise what work can be done, considering the project constraints.

The weekly plan is responsible for defining what will be done in the coming week, taking into consideration the goals achieved in during the previous week, the work scheduled in the look-ahead plan for that week, and the project constraints. Ongoing activities must be part of the inventory of executable work too. Figure 12.14 shows a sample of weekly schedule.

The weekly plan is developed in a meeting that takes place once a week (at the beginning or at the end of the week) with the last planners in attendance; this meeting should not be longer than two hours (Rodriguez et al., 2011). First, the last planners have to analyse the fulfilment of the previous weekly plan, detecting the causes of non-compliance so that they can make arrangements to correct the deficiencies. From this analysis of weekly compliance, changes can be made to the look-ahead plan. Learning is essential in the process; the systematic attack on the causes of non-compliance may increase the reliability of future planning. Then, the weekly meeting also sets the work that will be done in the coming week depending on the results of compliance of the previous plan. The weekly plan has to schedule only the work that is executable, once the existing restrictions are removed.

The reliability of the plan is measured in terms of percentage of plan completed (PPC), at the end of each week. The causes of non-compliance are analysed weekly in order to avoid them in the future. The reliability of the scheduling is directly related to productivity (González et al., 2008).

A basic support of the lean construction philosophy, which impregnates the LPS too, is the commitment of all participants (represented by the last

WEEKLY PLAN

PROJECT
Start Date
End Date

COD.	ACTIVITY	RESP.	GOAL			Fulfilment (yes/no)	Gantt Chart WEEK n					Causes of NON-Compliance							Description	Delay	Corrective Actions
			Expected	Completed	% Achieved		M	T	W	T	F	Suppliers	Subcontractors	Equipment	Safety & Health	Environment	External	Others			
							8	9	10	11	12										

Figure 12.14 Sample of weekly schedule (Rodriguez et al., 2011) (Reproduced with permission from Colegio de Ingenieros de Caminos, Canales y Puertos).

planners as final decision-makers) in the execution of the work scheduled. This commitment is reinforced by the public visibility of the results carried out weekly. This act of making public the weekly results of every party involved is an essential step to reinforcing the commitment of the past planners (Rodriguez et al., 2011).

Recent implementation experiences in different American countries show noteworthy achievements at the project level, as well as the diffusion of continuous improvement culture into the organisations. Summarisingly, LPS is a planning system in cascade whose main purpose, besides controlling the project, is to reduce the variability of the work by applying four basic principles (Rodriguez et al., 2009):

- personal commitment of the final decision-makers (last planners)
- coordination of the last planners through regular meetings
- use of a basic indicator of control called percentage of plan completed (PPC)
- public visibility of the results obtained weekly.

12.11   Time control

Project control analyses the partial results obtained, based on those estimated in the planning stage. The aim is to adopt possible corrective measures (Nicholas and Steyn, 2008). Time, cost and quality are examined individually and as a whole, usually in line with the following five steps for control displayed in Figure 12.3:

- acquisition of data
- measurement of work progress (time and cost) and performance
- comparison with the estimated indicator to detect deviations
- assessment of the causes which give rise to unforeseen events
- correction in order to redirect the deviation if negative, or to take advantage if positive.

Detailed planning usually reduces the number and the magnitude of the deviations (Turner, 2008). When these occur, the causes tend to stem from insufficient allocation of resources or technical reasons (e.g. greater complexity or the need to satisfy administrative timeframes which were previously not foreseen). The correction of these deviations may require negotiation with the client, particularly if such issues are attributable to an error in the preliminary information. Revision of planning, reallocation of resources or accepting the impossibility of a particular objective may be required, with their relevant contractual penalties. Usually the problematic activities are analysed; the purpose is to plan them with greater detail and factual data, in order to reallocate resources and meet the objectives in the best possible way.

This control process must be carried out frequently enough to be effective. The scope of the contract, resources used, complexity of the works and level of associated risk are typical factors which require greater frequency. However, the experiences of team members or longer execution periods require less insistence on control. Project control is carried out by considering regular periods (weeks or months normally) or milestones (e.g. invoicing). If a real time control system is available, this will be effective, assuming the data is entered on a regular basis (Harris et al., 2006).

Time is the easiest variable to control and as a result it tends to receive the most attention. If the project milestones are not reached at the proposed time or if the critical path is delayed, then the project will not be completed on time. The greatest difficulty is to establish clearly identifiable milestones (Nicholas and Steyn, 2008). To facilitate this limitation, the construction works can be subdivided into easily identifiable tasks and subtasks.

Time control is based on comparisons between the estimated and the real schedule. Information on real progress is compiled periodically (ideally every day or every week at the latest) and is used as an input in the control system. Mismatches are detected, comparing the estimated calendar with the real one. Once any deviations have been identified, they should be corrected as soon as possible. Solving differences requires corrective action such as rearrangement of resources or the use of bonuses for productivity on critical tasks (Pellicer et al., 2009).

12.12   Cost assessment and control

Project scheduling allows the cost to be calculated based on the resources assigned to each task. This is done based on the market prices for materials, equipment, machinery and manpower, combined with the performance of the teamwork involved in each activity (Harris et al., 2006). In order to complete a construction project sooner, we can accelerate or compress the activities of the critical path. However, this means that slack on non-critical activities will be reduced, and the chance of new critical paths occurring increases; hence, the risk of the project becoming late increases.

Once the project is awarded and before it begins, the contractor analyses all the work units (Winch, 2010). The economic study considers costs of material suppliers, subcontractors, manpower and company-owned machinery and equipment; it also takes into consideration the estimated performance of the teamwork, as obtained from the company's database or the experience of the field engineers. All of this is used by the company to determine the estimated budget, adding overhead costs and value added taxes to the previously computed costs. In this way, the global profit or loss can be estimated a priori in comparison with the contractual budget:

$$V = B - B'$$
$$B' = (\Sigma_j c_j \times m'_j) \times (1 + OVH) \times (1 + VAT)$$

where:

V	estimated profit or loss
B	contractual budget
B'	estimated budget
c_j	estimated cost of the activity
m'_j	estimated measure of the activity
j	estimated number of activities
OVH	contractor's overhead per unit
VAT	value added tax expressed per unit

The sum of direct and indirect costs is the execution budget for the project; this amount includes the expenses incurred on the construction site by the contractor. Direct costs are those incurred as part of the construction works and which can be directly attributed to a particular unit of the works: manpower, materials, machinery, equipment, tools and additional resources (Hinze, 2012). Manpower is involved directly in the execution of the work unit in question; for each unit the different labour categories involved must be determined, establishing the time employed and the cost per hour of work. For all material involved in each work unit, its amount used and its cost must be confirmed. Any additional materials which are not integrated into the work unit also have to be considered. These may be materials required for demolition, such as explosives. In the same way, any materials used to execute more than one work unit must also be analysed, such as formwork; this may be used on multiple events and it has a potential residual value. Furthermore, with regard to the machinery used in each unit, the time employed and the cost per hour of effective operation of each machine must be established.

Indirect costs are those incurred on site and which cannot be attributed to a particular work unit but to the project as a whole (Hinze, 2012). It is spread over all units as a percentage of direct costs. The following should be highlighted:

- technical personnel: construction site manager, field engineers, superintendent, foremen, surveyors, draftsmen and technical assistants
- administrative personnel: administrative chief and assistants
- construction facilities: jobsite offices to be used by the project manager and the contractor, warehouses, workshops, site fencing, access gates, parking facilities, etc.; the value of the buildings is estimated based on the surface area to be built

- other facilities and equipment: additional resources not included in the direct costs, general use vehicles assigned to the project, costs attributable to provisional energy, water and telephone facilities, etc.
- consumption: lighting, water, telephone, energy and fuel not included as a direct cost
- allowances, bonuses and expenses per kilometre to be paid for personnel transport, not included as a direct cost (the cost of direct labour transport is usually included in this section)
- various expenses: layout and abandonment, cleaning and maintenance during construction works, technical and legal advice, work meals, advertising, insurance, etc.

Economic control compares the real cost with that budgeted, analyses the difference and takes the necessary corrective measures. Every company has its own way of treating costs; some allocate them per activity, others by project, and there are even those who consider it based on the company as a whole. It is strongly recommended that costs are obtained per project, such that adequate control can be maintained over each contract (Harris et al., 2006).

Another important issue is the way that costs are obtained. Many companies use the general accounting system to calculate works costs; others use a specific system to control production (Pellicer et al., 2009). The difference between the two is the opportunity in analysing information. In conventional accounting there is normally a minimum delay of a month, in addition to the common delays in the payment and accounting assignment of subcontracting and specific supplies. Accounting does not provide the information in an adequate format either, as it is difficult to classify it per contracts.

The cost control system is important in allowing managers understand the current financial situation from the point of view of the company, the departments and the projects. In this way, it is possible to detect deviations of cost, time and resources, over those estimated, and to make appropriate decisions, the aim being to correct anomalies, optimise resources and fulfil the timeframe and budget for each project. In addition, adequate analyses of the data deliver conclusions regarding the productivity and profitability at a company, department, client or contract level. It allows timely detection of delays in invoicing, overtime and additional work not included under the contract, to name just a few (Pellicer et al., 2009).

Real-time control is only effective if all the parties involved enter daily data in line with the tasks carried out on each working day. The reliability of the data is fundamental, given that the expression 'rubbish in, rubbish out' can be applied perfectly to the system. If it is to be successful, all workers must have adequate awareness and good understanding of its benefits (Pellicer et al., 2009).

12.13   Earned value management

A technique commonly used for project planning and control is the earned value management method (EVM henceforth). EVM was introduced in the late 1960s in the USA by some public agencies (Flemming and Koppelman, 1996). It allows economic and time control of the project, considering the monetary repercussions that stem from a delay in the timeframe. Both time and cost variances on the proposed plan should be corrected as soon as possible, to ensure that the project can fulfil the original objectives.

Project managers perceive EVM as a powerful technique that gives early warning signals in order to plan and control projects in a proactive way (Abdul-Rahman et al. 2011). EVM it is widely used by public and private owners. The key benefits of EVM are (adapted from Fleming and Koppelman, 2006, and Abdul-Rahman et al., 2011): 1. integrated cost and time management; 2. improved vision of the project's scope; 3. early warning of problems; 4. foreseeing project deviation trends; 5. support for the decision-making process; and 6. the team's motivation to implement the project control process. Nevertheless, EVM in the construction industry is far behind other industries implementing the technique (De Marco and Narbaev 2013). In addition to the inherent characteristics of the construction industry, the main reason for this delay may be some limitations of EVM: 1. complex implementation; 2. deficient understanding of the technique; 3. distrust between owners and contractors; 4. pressures to report only profit; 6. contract payment scheme; 7. work scope changes and reworks; and 8. project size and duration.

To calculate time and cost variances, EVM considers inputs and outputs. Inputs are the periodically monitored actual expenditures as well as parameters (physical scope accomplishments). Outputs are the cost and schedule predictions (performance indices). The simplified terminology recommended by the Project Management Institute (PMI, 2008) is used throughout this book; the formulation is also the one proposed by the PMI (2008) and it is explained in depth in Flemming and Koppelman (2006). Three basic parameters are defined:

- budgeted cost of work scheduled (PV) or planned value
- budgeted cost of work performed (EV) or earned value
- actual cost of work performed (AC).

PV represents the originally budgeted cost based on the real performance. From the point of view of the contract, PV is the contractual budget less the proposed company profit. For a particular period, PV is calculated by adding the costs of each of the tasks completed and a proportional part of the tasks underway. The cumulative PV is the performance measurement baseline (PMB), whose value at the end of the planning process is called budget at completion (BAC). AC is the total cost actually incurred in accomplishing an activity; AC is measured during work execution. Finally, EV is the value of work performed expressed in terms of the approved budget assigned to an activity; this parameter is also measured during work execution.

Using the above definitions two kinds of variances can be computed: cost variance (CV) and schedule variance (SV). CV computes the cost performance linking physical performance to costs incurred. SV measures the schedule performance, showing whether or not the project is falling behind its baseline schedule. They can be formulated as follows (negative values indicate an excess over the budget):

- CV (cost variance)$= EV - AC$
- SV (schedule variance)$= EV - PV$

Both variances can be converted into percentages to compare different project performance and forecast the final project completion estimates:

- CVP (cost variance)$= (EV - AC) / EV$
- SVP (schedule variance)$= (EV - PV) / PV$

Performance indices can quantify the progress in terms of duration and cost:

- CPI (cost performance index)$= EV / AC$
- SPI (schedule performance index)$= EV / PV$

If the indices are equal to unity, the performance level is as planned. If they are greater than one, the performance is better than that planned; if they are less than one, the performance is below planned. These indices are normally used to predict tendencies and implement corrective action if necessary.

Normally, the accumulated cost over the course of the project's duration is drawn, displaying what is called an S curve (Figure 12.15). This name is due to the shape of the curve: it accelerates at the beginning of the project and decelerates at the end. It provides a visual representation of the project, showing whether costs are above or below the proposed levels. Normally, the curves are drawn in line with the budgeted cost of the planned work (PV), the budgeted cost of the performed work (EV) and the actual cost of the performed work (AC). This comparison shows if the project is fulfilling the planned cost and time standards.

As the project moves forwards, it shows not only what has been achieved, but also what needs to be done. This may lead to planning the project again, changing the completion date and the final cost. The estimated remaining cost to completion (ETC or 'estimated cost to complete the project') is computed as follows:

- BAC=budgeted cost at completion for the project (it is the PV at the expected date of completion)

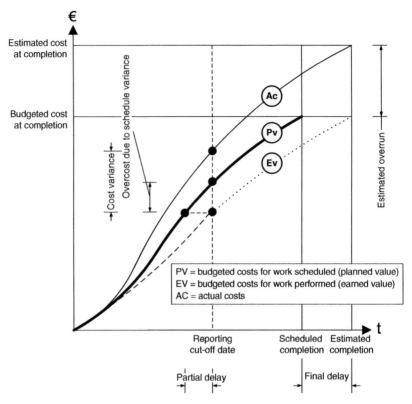

Figure 12.15   Graphic representation of the EVM method.

- EV = earned value (at the time of control)
- ETC = (BAC − EV) -1/ CPI

The estimated final cost (EAC or 'estimated cost at completion') is computed as:

- AC = actual cost (at the time of control)
- EAC = AC + ETC

Additional performance indicators can be used for estimating, such as earned duration or earned schedule, but the indices explained in this section are the most commonly used in practice. An in-depth explanation and discussion of these additional indicators and indices, as well a comparison with the typical ones, is given in Vanhoucke (2012).

12.14   Value engineering

Value engineering is the method used to identify and eliminate unwanted costs in a product, while improving the functional and quality requirements (Winch, 2010). The aim is to increase the value of products, and to supply them at the lowest possible prices. Its objective is to satisfy the product's performance requirements and the client's needs with the lowest possible cost. On a construction site, this involves: costs, availability of materials, construction methods, transport, limitations or restrictions of the site, planning and organisation, and analysis of costs and profits.

The concepts handled in value engineering are used in the work carried out by the project designers: architects or engineers. In fact, the project designer who does not consider the maximum economy in the selection and use of construction materials and methods is not being efficient in his/her job (Fisk and Reynolds, 2009). The value analysis of a product, service or process is more effective when an initial stage is designed, which can have an effect on the design, reducing costs and improving its performance. Some of the benefits obtained are the reduction in costs of the life-cycle, improvement in quality and reduction of environmental impacts.

However, the contractor can make the most of value engineering (Fisk and Reynolds, 2009). Therefore, the project designer selects the materials and construction procedures that can best adapt to the works after an analysis of the average market conditions, and the constructor knows the equipment, human resources and conditions that are best suited to complete the works. In some public works' tenders, the cost reductions and other improvements to construction procedures offered by the bidder are taken into account as a plus, so that the quality or functionality of the works is maintained at a high level.

The value engineering method includes the following aspects (adapted from Fisk and Reynolds, 2009):

- identify the main elements of a product, service or project
- analyse the functions of project elements
- develop alternative designs to execute these functions
- assess alternatives
- allocate costs to alternatives
- develop alternatives with a high potential for success.

The contractor, as well as the architect or engineer can, in the design phase, apply this method to improve construction procedures, which can produce significant cuts to the project's costs. These cost cuts can benefit the owner, who can award the construction project to the lowest bidder, while facilitating transparency and free competition between construction companies (Fisk and Reynolds, 2009).

The Chartered Institute of Building (CIOB, 2009) proposes the use of workshops or brainstorming forums (2–3 days) involving those who can contribute to the project or are interested parties at different times of the facility life-cycle: feasibility, design, procurement and construction. Participants should consider value as 'a function of cost and utility in its broadest sense'. Winch (2010), even though agreeing with this approach, states two key disadvantages: the (design) team can be demotivated because their proposals are dismissed or not considered worthy, and the time lost by the people attending the workshop plus the rework to be done as a consequence of the meeting, are additional costs for the owner.

CIOB (2009) also suggests that value techniques be used in conjunction with risk assessment techniques in order to make the best choices considering the risks at hand. Nonetheless, value engineering techniques have to be applied when there is a reasonable hope of cost savings and risk reduction, or when consensus is required but hard to accomplish. The next section will deal with risk management more deeply.

12.15   Risk management

Risk management deals with unpredictable events that might occur in the future, considering uncertain probability and consequence, and could potentially affect the objectives of the project (Loosemore et al., 2006). The event's outcomes can be positive or negative. Generally, the positive outcomes are called opportunities, whereas the term 'risk' refers only to the negative ones; this is the criterion that follows this section. From this point of view, risk management has to focus not only on minimising risks (or threats), but also on maximising opportunities at both project and business levels. Furthermore, as stated by Loosemore et al. (2006): 'every risk has an opportunity and every opportunity has a risk'.

Every construction project assumes a series of risks, since it is a new and unique activity, at least partially; it is said that 'project management is risk management' (Nicholas and Steyn, 2008). Nevertheless, many of these obstacles can be anticipated and controlled. The risk of the construction project grows significantly with uncertainty. Uncertainty is the main contributor to the risk of the project. Complete uncertainty arises because of a total lack of information, whereas certainty implies that the information is complete; between the two, runs the project. Decision-making is associated with the project itself and its inherent risk, involving reasonable assumptions (which are not infallible). It is very important to consider the interaction between risks and their possible magnification in order to avoid a domino effect.

Risk management is composed of four basic steps (Nicholas and Steyn, 2008; Turner, 2008): 1. Identify the risks; 2. Assess their probability and impact; 3. Develop strategies for reducing the risks; and 4. Monitor and control. Identifying risks means discovering what things could affect the project negatively. From this point of view, risk can be divided into internal and external (Nicholas and Steyn, 2008). Internal risks come directly from the lack of definition of the project, changes, extra work, technology, human resources or planning. They are controllable by the project manager (in a general sense), who can take steps to mitigate them. Their consequences are usually related to time delays, cost overruns and poor quality. External risks occur beyond the control of the project manager, although he/she can anticipate them and even influence them. These external risks may depend on, for example: the upper management, the labour market, taxation, inflation, social impact, political decisions, administrative paperwork, interpretation of the contract or natural disasters.

In order to identify the risks properly, some techniques can be used (Loosemore et al., 2006; Nicholas and Steyn, 2008): learn from past projects (see section 15.8 in Chapter 15), use checklists, analyse the work breakdown structure as well as the project schedule and use a cause-and-effect diagram or the Delphi technique (or any other variation of an expert panel). In some of these techniques, brainstorming could be applied as a tool for obtained rich feedback from the team or the stakeholders involved (Turner, 2008).

Once the possible risk factors and affected tasks are detected, the symptoms require special attention (Nicholas and Steyn, 2008). Symptoms are visible indicators that warn that the risk is materialising. From that moment on, contingencies or relevant protective actions must be implemented. For every risk, its probability, its impact on the cost and duration and its contingencies have to be considered.

Risk assessment aims to quantify the severity of the risk. This can be calculated by multiplying the probability of occurrence by the magnitude of loss/gain (impact). For these three factors, a scale rating is needed. For example, probability can be described as a percentage, a scale rating (e.g. from 0 to 1 or from 0 to 10) or an ordinal scale (e.g. low, medium, high). Each individual impact has to be assessed and given a rating, according to a scale in terms of cost impact, schedule impact or quality of the work performed. The severity of the risk can be ranked based on its probability of occurrence and its various possible impacts (Revere, 2003). Figure 12.16 proposes a simple and qualitative classification of the risk assessment according to their

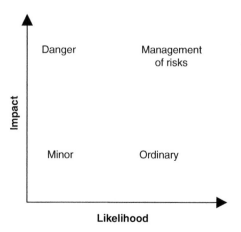

Figure 12.16   Risk assessment according to likelihood and impact (developed from Smith et al., 2006).

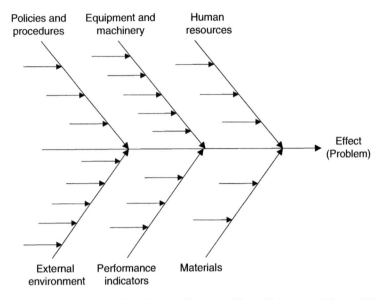

Policies and   Equipment and   Human
procedures   machinery   resources

Effect
(Problem)

External   Performance   Materials
environment   indicators

Figure 12.17   Cause-and-effect diagram (developed from Nicholas and Steyn, 2008).

likelihood (or probability) and impact that can be used as a first and intuitive approximation (Smith et al., 2006).

A cause-and-effect diagram, also called fishbone diagram or Ishikawa diagram (after its 1943 proposer), is a useful technique for analysing problems and finding solutions. Their graphic representation shows a horizontal axis where oblique lines arrive (fish bones) representing the categories of major causes of a problem; other perpendicular lines that meet the oblique ones represent the secondary causes (Nicholas and Steyn, 2008). Thus, the causes are grouped first by category (oblique lines), the most common being: people (physical or intellectual work), policies and procedures, equipment, materials, measurements and the external environment (Figure 12.17). Subsequent causes may be obtained through brainstorming sessions or using panels of experts. The root cause can be obtained by applying the technique of the five whys: it takes at least five rounds of questions to get the real answer, due to the fact the investigators have a tendency to stop at symptoms instead of going on to lower-level root causes. This technique was successfully applied in Toyota for many years, and now it is a key tool in the lean construction philosophy (section 11.8 in Chapter 11).

An event chart can be used to rank the assessment of risks developed previously. This can summarise the risks identified and assessed, thus allowing the project manager better decision-making (Revere, 2003). Figure 12.18 displays an example of a risk event chart (also called a risk register), which includes, for every risk: identity, type, likelihood (or probability), impact, severity and frequency. The greyish part of the table relates to Figure 12.16. Furthermore, the frequency column indicates if the risk event is one-time or repeated.

Risk	Type	Likelihood	Impact	Severity	Repetitive?

Figure 12.18   Risk event chart or risk register (developed from Revere, 2003).

Box 12.4   Different alternatives to address risks

1. Avoid the risk (ideal situation).
2. Divert all or part of the risk:
    a. transfer:
        i.   clauses of the contract (transferring it to the client)
        ii. clauses of subcontracts (transferring it to subcontractors)
        iii. insurance.
        iv. responsibility (the higher the responsibility regarding risks, the higher the authority on the project)
        v.   guarantees and warranties.
    b. Reduction:
        i.   use a competent and motivated team
        ii. hire experienced specialists
        iii. additional margins in the design regarding time and costs.
    c. Contingency fund:
        i.   economic ratio (increasing budget)
        ii. time buffer (increasing schedule): at the end of the project (applying the 'critical chain' method) or at the end of each task.
3. Accept risk: do nothing (generally for the low probability and impact cell in Figure 12.16).

Regarding the development of strategies for risk reduction, the first step is approaching feasible alternatives. Later, the optimal solution is chosen, an operation plan is proposed and, finally, it is implemented (Loosemore et al., 2006). The response can be to avoid the risk or deviate it totally or partially (transferring or solving it). This can be achieved with insurance policies, a guarantee or warranty, clauses in the main contract (transferring the risk to the client) or subcontracts (translating the risk to the subcontractors) or with a contingency fund (time and costs). The contingency fund should be divided into risk factors to prevent the unjustified and constant use of reserves. Box 12.4 (summarised from Nicholas and Steyn, 2008) shows the different alternatives for dealing with risks. Some specific risk reduction strategies that are focused on the design and construction phases of the facility life-cycle are explained in Revere (2003).

References

Abdul-Rahman, H., Wang, C. and Binti-Muhammad, N. (2011). Project performance monitoring methods used in Malaysia and perspectives of introducing EVA as a standard approach. *Journal of Civil Engineering and Management*, 17(3), 445–455.

Ahuja, H., Dozzi, S.P. and AbouRizk, S.M. (1995). *Project Management Techniques in Planning and Controlling Construction Projects* (2nd Edition). Wiley, New York.

Al Sarraj, Z. (1990). Formal development of line of balance technique. *Journal of Construction Engineering and Management*, 116(4), 689–704.

Arditi, D., Tokdemir, O.B. and Suh, K. (2001). Effect of learning on line-of-balance scheduling. *International Journal of Project Management*, 19(5), 265–277.

Ballard, G. and Howell, G.A. (1998). Shielding production: essential step in production control. *Journal of Construction Management and Engineering*, 124(1), 11–17.

Ballard, H.G. and Howell, G. (2003). Lean project management. *Building Research & Information*, 31(2), 119–133.

Burgess, A.R. and Killebrew, J.B. (1962). Variation in activity level on a cyclical arrow diagram. *Journal of Industrial Engineering*, 13, 76–83.

CIOB (2009). *Code of Practice for Project Management for Construction and Development*. Wiley-Blackwell, Oxford.

Clark, C.E. (1962). The PERT model for the distribution of an activity time. *Operations Research*, 10(3), 405–406.

De Marco A. and Narbaev T. (2013). Earned value-based performance monitoring of facility construction projects. *Journal of Facilities Management*, 11(1), 69–80.

Edum-Fotwe, F.T. and McCaffer, R. (2000). Developing project management competency: perspectives from the construction industry. *International Journal of Project Management*, 18(2), 111–124.

Fisk, E.R. and Reynolds, W.D. (2009). *Construction Project Administration* (9th Edition). Pearson, Upper Saddle River (New Jersey).

Fleming, Q.W. and Koppelman, J.M. (1996). *Earned Value Project Management*. Project Management Institute, Upper Darby (PA).

Fondahl, J.W. (1961). *A Non-Computer Approach to the Critical Path Method for the Construction Industry*. Technical Report No. 9. Department of Civil Engineering, Stanford University, Stanford (California).

Galloway, P.D. (2006). Survey of the construction industry relative to the use of CPM scheduling for construction projects. *Journal of Construction Engineering and Management*, 132(7), 697–711.

Goldratt, E.M. (1997). *Critical Chain*. The North River Press, Great Barrington (Massachusetts).

Goldratt, E.M. and Cox, J. (1984). *The Goal*. The North River Press, Great Barrington (Massachusetts).

González, V., Alarcón, L.F. and Mundaca, F. (2008). Investigating the relationship between planning reliability and project performance. *Production Planning and Control*, 19(5), 461–474.

Griffis, F.H. and Farr, J.V. (2000). *Construction Planning for Engineers*. McGraw-Hill, New York.

Harris, F., McCaffer, R. and Edum-Fotwe, F. (2006). *Modern Construction Management* (6th Edition). Blackwell, Oxford.

Hebert, J.E. and Deckro, R.F. (2011). Combining contemporary and traditional project management tools to resolve a project scheduling problem. *Computers & Operations Research*, 38(1), 21–32.

Hinze, J.W. (2012). *Construction Planning and Scheduling* (4th Edition). Prentice Hall, Upper Saddle River (New Jersey).

Kelley, J.E. (1961). Critical path planning and scheduling: mathematical basis. *Operations Research*, 9(3), 296–321.

Kerzner, H. (2009). *Project Management: A Systems Approach to Planning, Scheduling, and Controlling* (10th Edition). Wiley, New York.

Leach, L.P. (2000). *Critical Chain Project Management*. Artech House, Boston.

Loosemore, M., Raftery, J. and Reilly, C. (2006). *Risk Management in Projects* (2nd Edition). Taylor & Francis, New York.

Lu, M. and AbouRizk, S.M. (2000). Simplified CPM/PERT simulation model. *Journal of Construction Engineering and Management*, 126(3), 219–226.

Malcolm, D.G., Roseboom, J.H., Clark, C.E. and Fazar, W. (1959). Application of a technique for research and development program evaluation. *Operations Research*, 11(5), 646–669.

Mincks, W.R. and Johnston, H. (2010). *Construction Jobsite Management* (3rd Edition). Thomson, New York.

Nicholas, J.M. and Steyn, H. (2008). *Project Management for Business, Engineering, and Technology* (3rd Edition). Butterworth-Heinemann, Burlington (Massachusetts).

Packendorff, J. (1995). Inquiring into the temporary organization: new directions for project management research. *Scandinavian Journal of Management*, 11(4), 319–333.

Pellicer, E., Pellicer, T.M. and Catalá, J. (2009). An integrated control system for SMEs in the construction industry. *Revista de la Construcción*, 8(2), 4–17.

Pellicer, E. and Yepes, V. (2007). Gestión de recursos. In: *Organización y Gestión de Proyectos y Obras* (Martinez, G. and Pellicer, E. eds.), 13–44, McGraw-Hill, Madrid (in Spanish).

PMI (2008). *A Guide to the Project Management Body of Knowledge (PMBOK® Guide)* (4th Edition). Project Management Institute, Newtown Square (Pennsylvania).

Revere, J.J. (2003). *Construction Risk*. 1st Books Library, Bloomington (IN).

Rodríguez, A.D., Alarcón, L.F. and Pellicer, E. (2011). The management of the construction project from the perspective of the last planner. *Revista de Obras Públicas*, 3518, 35–44.

Smith, N.J., Merna, T. and Jobling, P. (2006). *Managing Risks in Construction Projects* (2nd Edition). Blackwell, Oxford.

Turner, J.R. (2008). *The Handbook of Project-based Management: Leading Strategic Change in Organizations*. McGraw Hill, New York.

Vanhoucke, M. (2012). *Project Management with Dynamic Scheduling*. Springer-Verlag, Berlin.

Wiest, J.D. (1964). Some properties of schedules for large projects with limited resources. *Operations Research*, 12(3), 395–418.

Wiest, J.D. and Levy, F.K. (1977). *A Management Guide to PERT/CPM*. Prentice Hall, Upper Saddle River, (New Jersey).

Wilson, J.M. (2003). Gantt charts: a centenary appreciation. *European Journal of Operational Research*, 149 (2), 430–437.

Winch, G.M. (2010). *Managing Construction Projects* (2nd Edition). Wiley, Oxford.

Further reading

Griffis, F.H. and Farr, J.V. (2000). *Construction Planning for Engineers*. McGraw-Hill, New York.

Hinze, J.W. (2012). *Construction Planning and Scheduling* (4th Edition). Prentice Hall, Upper Saddle River, (New Jersey).

Vanhoucke, M. (2012). *Project Management with Dynamic Scheduling*. Springer-Verlag, Berlin.

13 Progress Payment

13.1 Educational outcomes

The main objective of this chapter is to give information about progress payment under unit price contracting of construction projects. Other educational objectives are to:

- comprehend the basic definitions regarding progress payment
- explain the main types of contracts in the construction practice
- analyse unit price contracts in depth
- understand progress payment and payment processing procedures.

13.2 Introduction to progress payment

A progress payment is the disbursal of a portion of a construction loan after a certain stage of work has been completed. Actually, in their simplest form, progress payments may be viewed as a temporary interest-free loan from the owner (or the prime contractor) to the contractor (or subcontractor) until delivery of the product or service (under contract). They are extensively used in construction contracts because the deliverable usually takes a substantial amount of time to produce and consumes an amount of money greater than the contractor (or subcontractor) is able to finance with its own money. Therefore, progress payments aim at preventing the impairment of the contractor (or subcontractor) in performing or delivering a specific order by the owner.

Short construction contracts or those involving small amounts of money may possibly run without progress payments, but these become indispensable

Construction Management, First Edition. Eugenio Pellicer, Víctor Yepes, José C. Teixeira, Helder P. Moura and Joaquín Catalá.
© 2014 John Wiley & Sons, Ltd. Published 2014 by John Wiley & Sons, Ltd.

as larger sums of money are involved through longer construction periods. Accordingly, under many contract conditions, progress payments are optional for construction contracts of less than a certain value or duration, and are compulsory above those limits. In normal construction practice, small contracts tend to be paid in a single lump sum, typically at the end of the works, whereas interim payments are needed for larger contracts. This applies both to the owner/contractor and to the contractor/subcontractor relationship.

It should be noted that progress payments are not considered final payments in most construction regulations. They may be viewed as provisional or interim payments, which can only be retained by the contractor if it fully performs the contract and delivers all the items in accordance with the contract specifications (Ashworth, 2012). Alternatively, they may be considered partial payments if the construction contract can be structured in terms of incremental stages or deliveries and if there are appropriate acceptance criteria for the supplies, services or completed subsystems. In other words, interim payments are acceptable for the owner if partial units of the construction product can be safely inspected, tested and accepted without jeopardising their functioning as the final product still under construction. Otherwise, if the final product can only be successfully tested when it is fully built, it appears that interim payments are not a realistic alternative for the owner to consider. In this context, the client may be the owner or the main contractor when dealing with its subcontractors and other suppliers.

The construction contract starts the construction process. A wide variety of contract forms are used for construction projects, including a number of standard forms in several European countries (Ashworth, 2008; Cooke and Williams, 2009; Winch, 2010). Contract forms should be adapted to the procurement approach adopted for the project. The traditional procurement approach assumes the separation of design and construction, whereby the owner contracts the design first and then contracts the construction through an invitation or tendering procedure (Figure 13.1).

Traditional contract forms are as follows (Clough et al., 2005; Murdoch and Hughes, 2008; Winch, 2010; Ashworth, 2012): lump sum, unit price, cost plus, incentive and percentage of construction fee. As projects become larger and more complex so does procurement, including the following forms (see Chapter 3 for details): public-private partnership (PPP), public finance initiative (PFI), design-build-finance-operate (DBFO), build-own-operate-transfer (BOOT), plus – cost plus, reimbursable, target cost, cost plus fee and so on. This section briefly presents the traditional contract forms and further discusses the unit price contract procedure. Other forms of contract are more likely to have their own 'tailored' methods and formats for payments to the contractor or concessionaire.

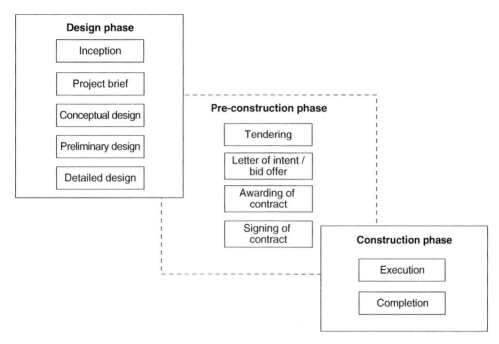

Figure 13.1   Traditional procurement approach.

13.3   Lump sum contract

Under a lump sum contract, the contractor agrees to perform the work described in the tender documents for a fixed sum of money, which covers construction costs, overhead and profit (Brook, 2008; Hinze, 2010; Winch, 2010). A lump sum or fixed-fee contract is suitable if the scope and schedule of the project are sufficiently defined to allow the contractor to estimate project costs. This means that the contractor has to gauge the project scope at the tender stage, taking into consideration the design documents and the description and quantities of works to be performed.

Measurements are normally included in the project specifications and assumed correct for tender purposes by contractors. Sometimes, measurements are not provided by private clients, requiring the contractor to determine work quantities for tendering purposes, but frequently declining liability for them. Alternative contractor design is sometimes possible at this stage, in which case the contractor is liable for the drawings and specifications substituted, in case of acceptance by the owner. Measurements are organised in bills of quantities, whereby works to be performed are broken down into bill items, often arranged in sections or chapters of the bill. Bills of quantities will be further discussed in the following section.

Under the lump sum rationale, the contractor is required to establish the total price for the project, irrespective of the actual quantities of work eventually performed on site. Because tender measurements are usually provided by the owner, the contractor needs to check them against the design documents and the site conditions, prior to establishing the fixed price for the project. Accordingly, most lump sum contract procedures allow the contractor to re-measure the works within a given time period after receiving the letter of intent from the owner. Typically, the outcome of this analysis made by the contractor is a claim for measurement errors for a set of work items of the contract bill and a claim for fresh work items missing in the tender bill. The contractor must use tender prices for the former and suggest new prices for the latter. The claims of the contractor are analysed by the owner and consultants, and the price for the project is finally established. Obviously, lump sum contracts are applicable when the design documents are clear about the scope and the quantity of works to undertake as described in the items of the bill of quantities.

13.4   Unit price contract

This kind of contract is based on estimated quantities of items included in the design documents, and their unit prices (Brook, 2008; Hinze, 2010; Winch, 2010). Work quantities are normally described in the project bill of quantities or schedule of construction works (Ashworth, 2012). The final price of the project is dependent on the quantities needed to carry out the construction works. It may be stated that a unit price contract is a contract in which the owner agrees to pay the contractor a specified amount of money for each unit of work successfully completed as set forth in the contract. Under the unit price logic, tender measurements are indicative of the amount of work to carry out on site. Therefore, the contractor gets paid according to the quantity of work actually performed, evaluated with the unit price stipulated in the corresponding item of the contract bill of quantities.

In general, a unit price contract is suitable if the scope of works, but not their quantities, can be accurately identified in the contract documents. This may happen either because design documents do not enable it or site conditions are unclear about it. For example, the amount of mortar needed for consolidating a wall is uncertain, so a lump sum approach would not be adequate for this work. Also, some contract arrangements allow the owner to release additional information on the project later than the tender, in which case this has to be clearly stated in the contract in terms of what and when.

Both lump sum and unit price contracts are frequently used in competitive tendering and it is not unusual to combine them in the same project, given that the first is applicable to the parts of the project where quantities may be clearly established and the second to those parts for which no quantities can be ascertained. The unit price and the lump sum contracts obviously correspond to different risk shares between the owner and the contractor. Concerning quantities of work to perform, the former places most of the risk at the contractor side whereas the latter balances risk distribution between them. But both may be used within the same contract, whereby works that can be clearly quantified are contracted under a lump sum approach and works for which such evaluation cannot be made are contracted under a unit price approach. This is often the case for foundations, which greatly depend on ground conditions.

Nevertheless, when compared to cost plus contracts, unit price contracts shift the production risk to the contractor, because for every work item performed, the contractor receives an amount computed on the basis of the unit price contracted, no matter how many difficulties and problems arise, which may increase the construction cost of that item. That is, the financial risks carried by the contractor are the difference between the actual and the budgeted cost to perform one unit of the work items. The risk to the owner is in the form of cost difference between the actual project cost and the estimated one as resulting from the tender bill quantities. For lump sum contracts, differences are settled before signing the contract after re-measurement is undertaken by the contractor, as signalled above, but for unit price contract, differences may become apparent much later than that and with more serious consequences.

But unit price contracts require careful preplanning by the owner or its consultants to select which work items the contractor should complete. Since the stipulated unit price is assembled at the tender stage by estimating the scope of works represented by the design documents released by the owner, any deviation from those documents will result in a change of scope, and the associated costs will be dealt with by a change order once the contract has been signed. The same applies to lump sum contracts. Therefore, the risk to the owner may be much higher than initially expected if contract documents are not accurately produced. For lump sum contracts, claims for change orders partially result from the re-measurement process after receiving the letter of intent as explained above, whereby fresh work items may be introduced if they are recognisably missed in the tender bill. But possibly, other missing works cannot be detected at this stage and will be revealed much later, during the construction phase.

From the above discussion, it may be concluded that both for lump sum and for unit price contract forms substantial care should be applied in organising the tender documents, normally requiring a comprehensive audit system.

The bill of quantities is a key document for unit price contracts because it allows for establishing the contract price, forms the basis for handling contract variations and is used as a cost control document by the owner. Bills of quantities are a product breakdown approach of the project into its constituent parts and they have been widely used in the European construction industry. Some countries have developed standards for measuring construction work, thereby easing dialogue between construction stakeholders and fostering the flow of cost information through the industry. These are two important arguments for the supporters of bills of quantities in construction.

However, bills have been criticised for their lack of relationship to the construction process on site. This is specially felt in building projects, for which a bill item may relate to works to be carried out in different locations and at different times. For other types of projects, for example road projects, bill items tend to be more similar to construction activities, thereby making bills of quantities much more process oriented. Another criticism of bills of quantities is that although they are relatively easy to work out from the complete design, they are much more difficult to establish in earlier stages of project development. For this reason, they are much more common in the traditional two-tier procurement approach.

But using a bill of quantities for cost management purposes has several advantages for both the owner and the contractor. The bill of quantities:

- allows for unit rate estimating, which is a rational approach to cost estimating
- enables the benchmarking of cost estimates against industry-level data sets
- creates a budget baseline for contracting and against which cost control may be easily performed during project execution.

The logic of the unit price is that the contract price is calculated by adding up the prices of each job described in the corresponding item of the bill of quantities. Item prices for tender purposes may be either calculated by the contractor or taken up from a benchmark in the industry. Clients often ascertain a tender baseline for limiting offers from the most qualified contractors in open tendering and for contractor guidance in the process of estimating the works. Alternatively, the owner may ask the contractor to stipulate a uniform multiplying factor applicable to the full range of prices. Factors above one lead to a tender price above the owner baseline and factors below one otherwise.

Box 13.1 shows a case study focused on the development of a system of standard technical specifications and bill of quantities.

Box 13.1   ProNIC, a case study of the development of standard technical specifications and bill of quantities

ProNIC is the Portuguese acronym for 'Protocol for the Standardisation of Technical Information in Construction' and refers to a research project that aims to develop and maintain a systematic and integrated set of credible technical contents, supported by a modern IT tool, to be used for reference by the Portuguese construction sector. The aims for developing ProNIC were the lack of available information for supporting technical decisions of project participants in Portugal, which is evidenced by: lack of technical information on resources required for construction work execution; dispersion of applicable standards, specifications and other technical documents; and inconsistent construction work description and specification contents. Therefore, technical documents produced by Portuguese construction professionals are disparate in content and substance and use various formats. The development of an adequate tool could help professionals in their duties and eventually improve conditions. Accordingly, the objectives of ProNIC are:

- to produce articulated standard construction specifications, particularly for buildings (new and refurbishment works) and roads

---

(Continued)

- to provide forms for works, materials and costs, associated with the specifications produced
- to generate all documentation needed to launch a contract, including budget estimates, detailed measurements and bills of quantities
- to allow for the comparison of contractors' offer bids.

The ProNIC tool comprises a database of knowledge of construction works and a set of software applications that enable the management and coordination of the contents, and their use by different stakeholders in the construction process.

13.5   Cost plus contract

This is a type of contract agreement wherein the owner agrees to pay the cost of all labour and materials, plus an amount for contractor overhead and profit, usually as a percentage of the labour and material cost (Clough et al., 2005; Hinze, 2010; Winch, 2010; Ashworth, 2012). The contracts may be specified as cost plus fixed: percentage, fee, fee with guaranteed maximum price, fee with bonus, fee with guaranteed maximum price and bonus, or fee with agreement for sharing any cost savings contract. This type of contract is favoured where the scope of the work is indeterminate or highly uncertain and the kinds of labour, material and equipment needed are also uncertain. Under this arrangement, complete records of all time and materials spent by the contractor on the work must be maintained. Table 13.1 explains how compensation is due to the contractor for each variant mentioned before.

Table 13.1   Types of cost plus contracts and compensation procedures.

Cost Plus Contract	Compensation
Cost+fixed percentage	Based on a percentage of the cost.
Cost+fixed fee	Based on a fixed sum independent of the final project cost. The owner agrees to reimburse the contractor's actual costs, regardless of the amount, and in addition to pay a negotiated fee independent of the amount of the actual costs.
Cost+fixed fee with guaranteed maximum price	Based on a fixed sum of money. The total project cost will not exceed an agreed upper limit.
Cost+fixed fee with bonus	Based on a fixed sum of money. A bonus is given if the project finishes below budget, ahead of schedule, etc.
Cost+fixed fee with guaranteed max price and bonus	Based on a fixed sum of money. The total project cost will not exceed an agreed upper limit and a bonus is given if the project is finished below budget, ahead of schedule, etc.
Cost+fixed fee with agreement for sharing any cost savings	Based on a fixed sum of money. Any cost savings are shared with the buyer and the contractor.

13.6 Incentive contract

Under incentive contracts, compensation is based on the engineering and/or contracting performance according an agreed target: cost, budget, schedule and/or performance. There are two basic categories of incentive contracts: fixed price incentive contracts and cost reimbursement incentive contracts. The fixed price incentive contract may be used when, although contract costs and performance requirements are reasonably certain, some cost uncertainty remains, particularity when the work is to be performed according to unique specifications and current market components are not usable. This is the case of projects involving new technologies or processes. Cost-based incentive contracts specify a maximum cost and a target profit for the contractor, and a target cost for the client. A formula is included to allocate any cost target over-run between the client and contractor: once the maximum cost has been reached, the contractor must subtract any additional overruns from his/her profit or fee, but if the project budget is not reached, the client's cost reduction is shared with the contractor.

The cost reimbursement contract provides the initially negotiated fee to be adjusted later by a formula based on the relationship of total allowable costs to total target costs. These contracts specify a target cost, minimum, maximum and target fees, and a fee adjustment formula. After project conclusion, the fee payable to the contractor by the client is determined in accordance with that formula.

Other incentive contracts may be tied to schedule rather than cost. Positive incentives can be built into the contract whereby a fee increase is due for early completion; conversely, negative incentives can also be included by reducing the payable fee if the schedule date is not met. This can apply to the final completion date of the project or be structured incrementally

---

Box 13.2 Awards for early compliance in the Portuguese code of public contracts

The Portuguese Code of Public Contracts allows the public client to award the contractor a compensation for early contract compliance, unless this is contrary to the law or the nature of the contract, according to the conditions established in the contract contrary to the law (article 301). This precept establishes the principle of free contractual stipulation regarding the award of compensation for early contract compliance, not defining the maximum amount for such awards; this is left to the public contracting entity that has the autonomy to establish it according to the specific circumstances. However, the precept precludes the possibility of assigning other awards, notably on grounds of the contract performance – or when the law or the nature of the contract do not allow it.

to a series of timeline milestones. Performance incentives relate to the achieving or exceeding of certain initial tight specifications (e.g. quality, safety or sustainability) for the given price. In that case, the contract grants a financial incentive as a compensation for investing more resources to meet those goals.

One of the drawbacks of these contracts is that they carry more administrative burden in order to maintain integrity, detailed cost accounting and performance tracking.

Box 13.2 shows the particular case of the Portuguese Code of Public Contracts.

13.7   Percentage of construction fee contract

This type is mostly used for engineering contracts. Construction costs are firstly estimated on the basis of resources and materials required for performing the works. Then, the client pays the contractor a fee that is estimated as a percentage of construction costs. The fee is intended to cover the contractor's overheads and profits and is agreed between the client and the contractor prior to the contract execution.

This contract is recommended when multiple changes to the job are anticipated, because it assures the contractor a fair return, while allowing for those changes to take place. However, it has various disadvantages: for the client because there is no guarantee of the final project cost – although this not only depends on the nature of the contract but also on the lack of accurate information on works to be performed; and for the contractor because costs and percentages may be difficult to explain to the client.

But over all, these contracts are considered poor business practice because the contractor has little incentive to hold down costs since its profit increases in proportion to costs incurred in performing the contract (a cost-plus-fixed-fee contract is a better approach). Accordingly, they are prohibited in some legal systems.

13.8   Progress payment procedures

For the reasons discussed in the preceding sections of this chapter, in construction practice, contracts between parties specify the methods of payment, including:

- the number of payments
- the amount of each payment
- the gap or stage of progress between payments
- the certification procedures for requested payments
- the date or stage when payment is due.

Also, contracts may specify approval and payment lags and set up retentions for guaranteeing the work performed or products delivered by contractors.

The most common methods of payment for traditional or design and build contracts are as follows:

- valuations: the traditional method of payment has been a physical measuring of the works carried out on site and the quantity of work evaluated against pre-established unit prices. Valuations are usually done monthly and are carried out jointly by the parties involved in the contract. The owner agent issues an interim certificate for the amount and the owner has to make the payment to the contractor within the period stated in the contract. The process is identical in a contractor/subcontractor trade
- milestones: the total contract value is broken down into a number of sums against predetermined milestones. Milestones are usually the completion of elements of the construction works (e.g. completion of the structure up to a certain level). A number of 20–40 milestones is common in construction practice. The acknowledgement by the owner agent that a milestone has been achieved will release payment of the sum to the contractor. This can sometimes be called an activity schedule (NEC contract) but has long been used for small construction projects in many European countries
- stage: this is similar to milestone payment but likely to have far fewer stages (e.g. completion of superstructure, achievement of watertight building)
- earned value: regular payments are made in accordance with an earned value system (see Chapter 12 for details on earned value management). Payment will be related to the actual progress position achieved on the works. As the value of payments is based on the schedule assessment of progress, it avoids the need to conduct separate monthly measurements of works carried out.

From the above discussion, it can be concluded that progress payments are made on the basis of the duly certified and approved status of the construction evolution. Status may be quantities of work performed, phases of the contract accomplished or scheduling assessment.

Although not formally recognised as a method of payment, special one-off payments may be provided by the owner or by the main contractor to its subcontractors or suppliers. This is the case of the advance payment, or simply the advance, that is the part of the contractually due sum that is paid in advance, while the balance will only follow after works, services or supplies are completed. An advance payment is requested by the contractor as a loan for financing materials or services needed for performing the works or ensuring supplies required by the contract. Another one-off payment, sometimes called ex-gratia, is a payment made in advance of the normal payment procedure to ensure that certain works are carried out in order to recover or prevent a delay situation or to expedite certain materials. Ex-gratia payments are also used in extreme cases when lack of cash is preventing the contractor from carrying out obligations under the contract. One-off payments are frequently accompanied by a prepayment insurance bond. It is important that, if payment is for materials or equipment, ownership is clearly established to guard against insolvency of the contractor.

Approval and payment time-lags are common in most contract conditions and follow the typical progress payment procedures. Although lags are typical of the construction activity, it must be remembered at the time of designing contract conditions that timely and accurate payment of the contractor for completed contract work and completed change order work is a critical element of successful cost control.

First, the contractor (or subcontractor) gets certification and approval by the owner agent (or the main contractor) for an estimate of the work performed. Under a valuation approach, a date is jointly agreed for carrying out the valuation of works done during the previous billing period (typically a month, but other periods may be established in the contract). Normally, a day of the month is established in the contract for the contractor to submit to the owner the estimate of work performed, thus a previous date for valuation must be settled. In the relationship between the main contractor and its subcontractors and suppliers, compatible dates ought to be established. Under this approach, an interim certificate for the amount due is issued by the owner agent (or by the main contractor representative) in a short period of time, compatible with the date for submission of the estimate.

But in some contractual arrangements, it is the contractor's initiative to submit the estimate of the work performed to the owner without previous certification. The owner designates a person for revising, certifying and approval of the estimate within a pre-established period of time. The estimate is deemed approved and certified after that time, unless before that time the owner issues a specific written finding detailing those items in the estimate of the work that are not approved and certified.

Under a milestone or stage method of payment, the contractor (or the subcontractor) asks the owner agent (or the main contractor representative) to acknowledge that a milestone or stage has been achieved. The corresponding construction status is deemed achieved after a pre-established period of time, unless before that time the owner issues a written finding justifying the opposite.

If the method of payment is based on the earned value, the schedule assessment of progress is demanded and acknowledged following an identical process to the milestone or stage method described above.

Finally, the progress payment is released by the owner to the contractor or by the main contractor to its subcontractors and suppliers before a time period established in the contract. Thirty-day periods are common but there are cases of sixty-day or longer periods in the European construction industry nowadays. If any progress payment or a part of it is delayed after the date due, interest is paid at an agreed rate on such unpaid balances as may be due.

In most contract conditions, the owner does not release the full progress payment but withholds an amount for guaranteeing the work performed or products delivered by contractors. Retaining a percentage of the progress payment is the most common process followed by owners (Hinze, 2010), values of 4–10% being typical. Alternatively, the contractor may provide a substitute security bond for the retention established in the contract. The same applies to any entity in the supply chain, provided that the retention does not exceed the actual percentage retained by owner of that entity, unless otherwise stipulated in the contract between the two parties involved.

Furthermore, the owner or any entity in the supply chain may withhold payment to the subcontractor or material supplier for unsatisfactory job progress, defective construction work or materials not remedied, disputed work or materials, third-party claims, failure of a subcontractor to make timely payments for labour, equipment and materials, damage to the contractor or another subcontractor, or reasonable evidence that the subcontract cannot be completed for the unpaid balance of previous retentions.

With the aim of preserving the construction supply chain, some contract conditions include a set of items guaranteeing the cascade payments to the subcontractors and suppliers of the main contractor after each progress payment is released by the owner. Conditions may establish that the main contractor must pay its subcontractors and material suppliers (and each subcontractor must pay to the subcontractor's subcontractor or material supplier), within a given time period after receiving the progress payment, unless otherwise agreed by the parties. Typical payment lags are of a week or so, unless otherwise stipulated, and after that period the subcontractor or material supplier may notify the owner of any payment less than the amount or percentage approved. As a means of preventing the lack of payment from the main contractor, subcontractors may request the owner to be notified of any progress payment made to the main contractor in the course of the involvement of the subcontractor in the project. Despite the purpose of protecting the construction supply chain from the lack of payment from the main contractor, the above premises seldom apply to amounts payable for design services, pre-construction services, financial services, maintenance services, operations services or any other related services included in the contract.

References

Ashworth, A. (2008). *Precontract Studies. Development Economics, Tendering and Estimating* (3th Edition). Blackwell, Oxford.

Ashworth, A. (2012). *Contractual Procedures in the Construction Industry* (6th Edition). Pearson, New York.

Brook, M. (2008). *Estimating and Tendering for Construction Work* (4th Edition). Butterworth-Heinemann, Oxford.

Clough, R., Sears, G. and Sears, K. (2005). *Construction Contracting: A Practical Guide to Company Management* (7th Edition). Wiley, New York.

Cooke, B. and Williams, P. (2009). *Construction Planning, Programming and Control* (3rd Edition). Wiley, Oxford.

Hinze, J. (2010). *Construction Contracts* (3rd Edition). McGraw-Hill, New York.

Murdoch, J. and Hughes, W. (2008). *Construction Contracts: Law and Management* (4th Edition). Taylor & Francis, London.

Winch, G.M. (2010). *Managing Construction Projects* (2nd Edition). Wiley, Oxford.

Further reading

Ashworth, A. (2012). *Contractual Procedures in the Construction Industry* (6th Edition). Pearson, New York.

Hinze, J. (2010). *Construction Contracts* (3rd Edition). McGraw-Hill, New York.

14 Claims and Change Management

14.1 Educational outcomes

The aim of this chapter is to give readers a practical guide about adequately managing, preparing and negotiating claims and change orders arising from the implementation of construction projects, and how they can be recognised early, in order to avoid further costly and timely litigation. By the end of this chapter readers will be able to:

- comprehend the main concepts related to claims and change management in construction
- identify and analyse the causes of claims
- determine and classify different types of claims
- understand the claim management process
- implement conflict avoidance practices
- evaluate and correctly deal with change management process.

14.2 Introduction to claims and change management

Dealing with claims is an unavoidable feature of the construction industry, absorbing precious time, money and staff resources (Fisk and Reynolds, 2009). In fact, construction contracts are commercial relationships established between two main parties: the clients and the contractors. And, as happens in other economic activities, to stay in business, firms must have profits. However, owing to the complexity of construction processes, competitive bidding and the risks inherent in construction activities, the way

Construction Management, First Edition. Eugenio Pellicer, Víctor Yepes, José C. Teixeira, Helder P. Moura and Joaquín Catalá.
© 2014 John Wiley & Sons, Ltd. Published 2014 by John Wiley & Sons, Ltd.

things were planned sometimes cannot be done. There are many reasons for this – some well known: the conditions on the site differ from those expected and represented on the design; additional work and change orders issued by owners; defective contract documents; abnormal weather conditions; unexpected escalation of material and labour costs.

Depending to an extent on the type of contracting system, it is expected that contractors will present a claim in order to get compensated with additional payment or time extension, if the type of incidents referred to above arise during construction. Sometimes this procedure is precisely the administrative process required to handle that type of construction events, and get fair and timely compensation for additional costs incurred (Kutner, 1989). So it may be stated that, despite mainstream opinion, the presentation of claims should not be interpreted as an indication of either a legalistic attitude or mismanagement or devious practices; it is just an integral part of the construction process, a safety valve and a contractor's right to be reimbursed for losses and extra costs incurred. Even successful projects experience construction claims, and good claims administration and management are as important as good engineering, safety or business principles (Levin, 1998).

14.3 Definition of claim

The definition of a claim depends on what is established in the conditions of contract used for the construction works. Sometimes there is no explicit definition, but just references to the type of event connected with claims: for example, extension of time, defective documents, change orders or *force majeure* events (Powell-Smith et al., 1999). For the purpose of this chapter, the following definition of construction claim can be used: written demand by one contracting party, seeking, as matter of legal right, the payment of additional money, the adjustment of the interpretation of contract terms, an extension of time to perform the contract, or any other relief arising under or related to a given construction contract (Box 14.1). However, during the normal development of a contract, which aim is to repair, maintain or build a facility, different situations occur that can be defined as:

- problems: are parts of the normal process of a construction project and should be solved on a daily basis within the project team
- disagreements: can be defined as problems that can only be solved after hard and substantial negotiations between the contracting parties
- disputes: occur when the project team is unable to solve the disagreements, and persons outside construction site offices become involved in the process
- conflicts: take place when external expertise is needed to get a solution to the dispute, in a timely and fair manner
- litigation: is a process that assumes a legal form, such as a lawsuit or a petition, and needs to be solved by courts or other legal forums.

Box 14.1   Castro Daire bypass motorway

To get a high capacity access to the interior north of Portugal, avoid crossing urban areas by heavy traffic, and bypass the mountainous territory of the region, the Portuguese Road Department launched several construction projects, including a local motorway with 16 km and 2×2 lanes cross-section, with an extra lane for low speed vehicles for 3 km, a tunnel of 820 m length and three special bridges over the rivers Paiva and Paivô.

The total investment reached €70 M, including land acquisitions. In respect to the motorway, the tender price was about €26.4 M and the expected duration was two years, subject to a premium price in case of early finish. However, due to different issues that occurred in the construction phase, namely abnormal weather, major design changes including a new intersection that affected the equilibrium of earthworks, and lack of productivity, it actually cost €36 M, and suffered a delay of five months.

These reasons led the contractor to present three different claims: 1. a global claim of €7.5 M due to the design change orders issued by the client, including lost productivity, additional direct costs and overheads; 2. a claim for abnormal weather of €2.0 M; and 3. a claim related to the price award for early finishing, requesting the maximum amount of 10% over the final cost of the project (€3.2 M).

The resolution of the claims followed different procedures. The first was discussed between the parties, with the participation of an external consultant, and an agreement was reached, since the client accepted that the design changes caused disruption and delays in the approved schedule. Consequently, an additional amount of €3.4 M was paid to contractor. The *force majeure* event was refused due to lack of substantiation, and the contractor did not appeal. The final claim went through a mediation process and the conclusion was that the right to collect the maximum premium subsisted, even though the works finished later than initially foreseen, because delays were excusable from the contractors' point of view. However, the percentage only applies over the initial tender price, since contractor could not, at the time of the bid, anticipate cost overruns. Consequently, the claim was settled at €2.6 M.

Due to the lack of precision and the wide variety of standard forms of contractual conditions, the words claim, conflict and dispute, are used, together or separately, sometimes indistinctly, and without clear indication of their precise meaning. In order to establish the differences between claim, dispute and conflict, some authors suggest an evolution process showing the

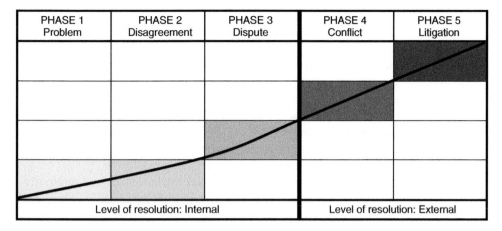

PHASE 1 Problem	PHASE 2 Disagreement	PHASE 3 Dispute	PHASE 4 Conflict	PHASE 5 Litigation

Level of resolution: Internal   Level of resolution: External

Figure 14.1   Process evolution of claims in construction (developed from Bramble and Cippollini, 1995).

different semantics (Moura and Teixeira, 2010; Kumaraswamy, 1997). Accordingly, the claim process evolves in five distinct stages of relationship between the claimant (the contractor) and the client; the first three phases allow for resolution levels within the project scope and the others engage external help to aid resolution, as shown in Figure 14.1.

14.4   Causes of claims

It is generally recognised that the environment of the construction industry, owing to the highly complex nature of its activities and the interrelation with a multitude of factors, is particularly vulnerable to litigation (Moura and Teixeira, 2007). In fact, there is no other industry which, after an intense competitive bidding, produces an extremely complex product, never built before, using extensive working hours of manpower and equipment for a fixed price, within predefined time limits, by a construction team who had probably never worked together previously.

While the number of construction claims increases and their preparation becomes more sophisticated, the issues are broad and are extensively reported in the international literature. At a project level, they include design errors, client change orders, project scope increases, abnormal weather conditions, restricted access to site, acceleration, defective contract documents, different site conditions, scheduling problems, design defects and third-party actions/inactions (Moura and Teixeira, 2007).

Nevertheless, some authors have found ways of classifying the underlying causes for the presentation of claims. For example, from the perspective of the construction industry and its role among other economic activities, the reasons that lead to construction claims can be divided into (Semple et al., 1994):

- economic reasons such as recession, inflation and unemployment, as well as political reasons, such as the decreasing of public investment or the redefinition of priorities
- market reasons, such as internal and external concurrency, the competitive bidding process and risk allowances in bids
- contractual reasons, such as the complexity of the construction process, new forms of contracts, biased risk allocation and exculpatory clauses in the contract documents
- legal reasons, for instance the new lawsuits on environment protection, health and safety responsibilities and funding restrictions
- organisational reasons related to complex and bureaucratic administrative procedures, inadequate empowerment or inaccurate quality control during the design phase.

Other types of categorisation are used to measure the construction contract dispute predictability, such as the analysis of a proposed hierarchy of contract dispute predictors constructed around three main categories related to people, project and processes (Diekmann and Girard, 1995), as described in Figure 14.2.

PEOPLE:		
Client:	**Contractor:**	**Business relationships:**
• Capable management	• Capable management	• Team building
• Effectiveness of responsibility	• Effectiveness of responsibility	• History together
• Experience with type of project	• Experience with type of project	• Power balance
• Success of past projects	• Success of past projects	• Expectations of further work
• Experience and competence	• Experience and competence	
• Motivation	• Motivation	
• Interpersonal skills	• Interpersonal skills	
PROJECT:		
External:	**Internal:**	
• Environmental issues	• Pioneer project	
• Project interference	• Design complexity	
• Site limitations	• Construction complexity	
• Remoteness	• Size	
PROCESS:		
Pre-construction stage:	**Construction contract:**	
• Input from all groups involved	• Realistic obligations	
• Financial planning	• Risk identification and allocation	
• Permits and regulations	• Adequacy of technical plans and specifications	
• Scope definitions	• Formal dispute resolution process	
	• Operating procedures	

Figure 14.2   People, project and process categories of claims indicators (summarised from Diekmann and Girard, 1995).

In fact, issues involving people and the way each participant in the construction process deals with potential conflicts are extremely important as they also affect organisations and the relationships between them. Project issues are those characteristics that define the technical nature of the construction work to be performed, such as its type, size or complexity. Finally, the process issues are those related to the way that the construction contract and the building process is carried out. This categorisation is an indicator of whether a particular construction project should experience claims and disputes during its implementation phase.

Besides the above classification, researchers and practitioners have developed several surveys that aim to find the root causes for the presentation of construction claims, independently of the legal system or the form of contract. For example, Zaneldin (2006) ranked by frequency 26 causes for claims presented in several construction projects, and found that the 15 most important causes were: change orders, delay caused by client including in payments, low price of the contract due to bid competition, changes in materials and labour cost, client personnel character, variation in estimated quantities, subcontracting problems, delays caused by contractors, poor contractor management, contractor financial problems, deficient quality performance, government regulations, estimating errors, scheduling errors, design errors and omissions.

Kumaraswamy (1997) identifies the main causes for claims as:

- time claim: inclement weather, hoisting of storm signal, instruction issued to resolve discrepancy, variation order, substantial increase in quantity of a work item, delayed possession of site, disruption of regular progress, suspension of the work, delay caused by a third party
- cost claim: ambiguity in documents, construction method change imposed by client, facilities provided to another contractor in the site, additional tests, uncovering of works for examination, delayed possession of site, acceleration of works, suspension of the work, additional works arising from repair/defects, interest on claims due to late payment, disruption to regular progress, client's breach of contract, changes and variations.

Despite the various causes that can contribute to the presentation of construction claims, the important issue here is the correct identification of the problem, in order to adequately prevent, or document and quantify, the damages to claim. For instance, in case of an inclement weather claim, the contractor should be aware of the relations between the manpower productivity and the environmental factors (temperature, humidity, noise, wind, precipitation, etc.) in order to correctly demonstrate and evaluate the related loss of productivity (Adrian, 1993; Horner and Talhouni, 1996).

14.5   Types of claims

To some extent every claim has its unique characteristics in relation to the causes of occurrence, the liability issues, the calculation of damages and the form of presentation. So it is likely that there are as many specific reasons for the occurrence of construction claims as there are different types of claims. Nevertheless, despite the different contracting systems and standard contract conditions, claims essentially derive from construction events. And those events can be grouped into categories. The aim of grouping claim events in categories is to better prepare the presentation or the defence against a claim. In fact, the early recognition of a claim, a fundamental step in the management process, is the key issue to winning (or preventing) a claim as it enables to isolate it properly and better identify and evaluate the claim's damages.

Although the best practices point to the categorisation of construction into different types, it is common to see contractors still presenting global claims, meaning that the amount claimed corresponds to the total costs supported by the contractor, added to the expect profit, and discounted from the payments received from the client. However, most of the contractual specifications and construction legal frames do not accept this type of claim, because the standards of substantiation oblige contractors (or whoever presents the claim) to prove that (Davison, 2003):

- some particular (unpredicted) event occurred
- according to contract conditions or applied law principles, this event (or the absence of it) implicates liability of the counterpart of the contract (cause–effect)
- the event, direct or indirectly, caused loss or damages.

The claimant should be able to demonstrate that, for instance, the actual work scope, the change order issued or the work condition is somehow different from what could reasonably be expected at the time of the bidding. Taking in account the uniformity of the contractual language and the advantages in early recognition and identification of a claim event, it is frequent to classify construction claims according to the event that initiated the claim or the contract clause that allowed liability. Even though depending on different contractual systems, several studies show that the most frequent types of claims are delays (including those due to abnormal weather conditions) and change orders either direct or indirect (which includes different site conditions and defective contract documents) (Moura and Teixeira, 2010).

Change claim occurs whenever a client issues a change order to the contractor, not just to execute some extra works, but also to change the planned schedule or to increase or reduce the project's scope. This type of claim can be divided into two main subtypes: direct change orders and indirect changes, the latter resulting from an action or inaction of the client that induces some kind of loss to the contractor (time or money), where the client is not aware of or does not recognise it as a change. Common examples of indirect change orders that occur on every construction site are: defective specifications and contradictory contract document clauses, different interpretation of contract provisions, abnormal or abusive behaviour during site inspection of the work performed, design errors, implied duties and improper rejection of construction materials. As a subtype of the direct change claim, the different site conditions claim is one of the most invoked events to claim additional costs, and it occurs whenever ground, soil or other geotechnical variable is different from what was represented or was expected, according to contract documents.

Together with the changes claim, the delay type is also one of those presented by contractors, as they request not only just the payment of extra money, but also time extensions to perform the contract. Normally, a delay claim arises when a contractor is not able to complete the project within the planned time, for whatever reasons, given that these reasons are not his/her responsibility. If the contractor stays on the site for additional time beyond the schedule, substantial extra costs can appear that ought to be claimed, including not just direct costs but also overheads and opportunity costs.

Suspensions of works, acceleration, payments, termination and *force majeure* events, complete the main different type of claims that can be presented in a construction project; although there are studies that identify others, they can be considered as subtypes of the main types depicted here (Moura and Teixeira, 2007).

14.6   Claim management process

Construction claims are an integral part of the construction process and cannot be avoided. Sometimes, they are the only way for the contractor to achieve fair and timely compensation for additional costs incurred due to a variety of situations unforeseen at the bid time and not stipulated on the contractual bill of quantities. To manage the claim process adequately it is necessary to undergo five different stages (Kululanga, 2001), some of them overlapping in time (Figure 14.3): identification, notification, documentation and quantification, presentation, and resolution.

The first step, and perhaps the most important, is the claim identification, which involves timely and accurate claim prediction. This can be done after the occurrence (or the absence) of some event, or by inspecting contract documents. Note that, frequently, what seems to be a minor issue without importance to the daily routine on the construction site, can suddenly change into a costly claim, leaving its resolution to outside the project level. Field staff seldom have enough experience and skills to perform this task correctly, and often need assistance, in order to identify a potential claim event in a timely way (Box 14.2 developed from the Portuguese Auditing Court, report 37/2008).

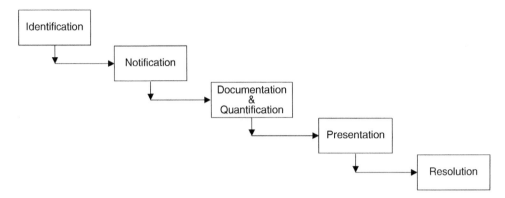

Figure 14.3   Claim management process (developed from Kululanga et al., 2001).

Box 14.2   Porto Casa da Música

The Porto Casa da Musica is nowadays a symbol of the city, located in its financial heart (Boavista neighbourhood), and an architectural master-piece designed by the world-renowned architect Rem Koolhaas. Casa da Música was conceived to mark 2001, the year in which Porto was Cultural Capital of Europe, and it is the first new building in Portugal entirely dedicated to music, including presentation to the public, music education and composition. The project began in 1999 after an international architectural competition, and opened its doors to the public on April, 2005. Casa da Música was part of the urban redevelopment of Porto and was integrated into a network of cultural facilities, for the city and for the wider world, and it was planned to reach all types of public. It is an innovative and wide-ranging cultural project, which aims to make an exciting contribution to the national and international music scene, as an arena for all types of musical events – from classical music to jazz, from 'fado' to electronic music, from great international productions to experimental projects.

It is also a complex and spectacular building, constructed in white concrete with the shape of a three-dimensional polygon, inside an exterior shell made of sloping walls and piles; $22\,000\,m^2$ of construction are distributed onto twelve floors, three of them underground, including a main audit with 1300 seats, and a secondary auditorium for 650, both with special provisions due to acoustic quality assurance, several rehearsal rooms, recording studios, documentation centre, library, multipurpose foyer for exhibitions for 2000 people, VIP room, dressing rooms, a restaurant at the top, several bars, classrooms, technical areas and parking for 700 cars.

(Continued)

However, during the construction phase, it experienced several claims, delays and cost overruns. In fact, a facility that was programmed to be finished in 2001, to be constructed in 28 months, with an estimated costs of €33 M, was only opened to the public after almost seven years of construction work, suffering an accumulated delay of 4½ years (193%), and with a final cost of €111 M (228% cost overrun). One of the most discussed items was the decision over a claim presented by the contractor alleging that the shoring and formwork for the sloping concrete walls were not included in the bid price. In fact, it was necessary to design two independent special structures: one vertical and metallic, that supported the secondary; the other one sloped, made of wood, which supported the concrete forms. The client argued that the means and the methods to do the work were the contractor's responsibility, so the tender should include the price to execute the building with its designed forms.

Nevertheless, after a mediation process, and just before the arbitral court intervention, the client decided to pay €3.6 M extra to the contractor, as well as granting 18 months of extra time, recognising that the unusual form and the slope of the building in which stability was only guaranteed after the conclusion of all exterior structure, needed a temporary support for the form, whose cost could not be properly anticipated at the time of the bid.

The following step is the claim notification. According to most standard contractual conditions, the claim notification involves the information addressed to the counterpart that a potential problem is arising. Normally, this must be done in writing, within a few days of the occurrence of the event, depending on the type of claim issued (Powell-Smith et al., 1999). Very often a contractor does not know the real cause of a claim situation, until some time after the occurrence of the event. Nevertheless, prudence advises that he/she gives the earliest possible warning to the owner, in order to preserve the rights to a future claim, after the time needed for further information to be collected and analysed. The other party, normally the owner, should immediately face the problem by holding a meeting where the parties discuss the situation, and its implications for the contract time and budget.

Then, the burden of proof implies getting the necessary claim documentation. This involves the collection of all factual grounds on which the claim is based, in order to prove that some particular event occurred and caused some kind of damages to the contractor. For that purpose, a set of records are useful and should include: tender and contract documents; orders and instructions to the contractor; general correspondence with the client; request for information; subcontractors quotations and contracts; vendors purchase orders and deliveries; daily records of manpower and equipment; daily production logs; productivity estimates; accounting and cost control records; task progression schedules and updates; progress payments; progress reports; time-lapse photographs and videos; minutes of meetings; site inspections; quality control inspections; safety reports and so on (Rubin et al., 1999). Also, according to the principles established in most legal systems, the claim's legal basis must be found, that is the specific clause of the contract or the common law principle conferring the right of being reimbursed must be invoked.

The other part of this step is claim quantification, which includes the use of costing and scheduling methods recognised by the construction industry for the calculation of the extra time and money claimed. The claimant should give a substantive and detailed description of the extra costs incurred, or to be incurred, because of the claim situation, which must be based on actual recorded costs, on reasonable costs derived from existing project data (e.g. tender prices) or on accurately estimated costs. The items normally included in the pricing of a claim are: direct costs of materials, equipments and labour; additional subcontractor or vendor costs; additional stocking costs; lost productivity costs; other delay costs (escalation or standby time); acceleration costs (overtime premium or disruption); jobsite and home office overheads; additional bonds and insurance costs; interest or finance costs; lost profit; loss of opportunity costs and profit; legal and expertise fees and claim preparation costs (in case of litigation).

The next step is the claim presentation, which is the act of preparing the formal document and sending it to the other party; Box 14.3 describes best practices that should be used for the preparation of the document. This document must be well organised, logically built up, clearly written and in a format that leaves no room for doubts about the reasons that contributed to the claim submission (Carnell, 2000). It can be divided into two main parts: the right to the entitlement, more descriptive and based on contract clauses; and the quantification, based on accountability, cost evaluation and scheduling allowed principles, which must contain all the necessary documentation or invoked scientific methodologies, allowing for proper analysis. The claimant must be aware that this document may be discussed in court, if no agreement is reached.

The final part of the claim management process is the claim resolution, the objective of which is to settle an agreement satisfying both parties as quickly as possible. This settlement may be reached in one of several ways. The first, and most inexpensive, is by direct negotiation between the parts. Normally, negotiations begin with the initial meeting where participants establish the procedures to be followed in succeeding steps, state their positions, highlight the main areas of agreement and disagreement, and settle what type of records or data will be accepted. Another important issue for this meeting is that it enables each party to know the other's strategy and attitude. So team members should: be well prepared; have previously established the boundaries of possible settlements; be flexible; avoid personal conflicts; and adopt correct negotiation techniques such as win-win approaches. In order to achieve

---

Box 14.3   Preparation of the claim document

The basis of a successful claim is a good, reliable and consistent document that will be read, analysed and sometimes decided on by a skilled practitioner in the construction field. The table of contents for a claim document should be:

1. **Package:**
   a. Cover with description of the project and the claim
   b. Letter of presentation
   c. Table of contents and table of figures
   d. Description of the contracting parties
2. **Body of the claim:**
   a. Summary
   b. Main facts and findings: brief description of the project and exhaustive description of the disputed issue (chronology of facts, not opinions)
   c. Analysis of applicable contract terms, breach of contract or general and particular provisions, drawings, contract law, other cases, court decisions, etc.
   d. Description of the original strategy at the time of the tender, foreseeable schedule, expected cost and profit and general expectations
   e. The relation cause–effect: the type of the claim and the consequence (the actual progress, the mitigation actions and consequences)
   f. The damages and additional costs incurred: direct costs, overheads, productivity losses, stocking, escalation, opportunity costs and lost profit
   g. Overall analysis and conclusions
3. **Appendices and annexes:**
   a. Source documents: daily logs, minutes of meetings, material catalogues, relevant tender documents, vendor quotes, shop drawings, measurement and payments, health/safety or quality audits, change orders, letters, etc.
   b. Prepared exhibits, presentations, photos and videos
   c. Detailed schedule analysis, summary of delays and schedule impacts
   d. Detailed calculations of the damages (point 2 F)

---

success, both parties should avoid adversarial attitudes and prepare the meetings carefully, ascertaining all the relevant information, avoiding meaningless arguments, concentrating on the claim's scope, using the other party's weaknesses and anticipating their arguments.

If negotiations between parties reach an impasse, it is necessary to bring in outside help. The next step is mediation, which is a process for settling disputes through conferences conducted by a neutral third party who facilitates negotiation (Carnell, 2000). The role of this mediator is to cultivate empathy and understanding between the parties, overcoming barriers while acting as a confidant, tension breaker and a communication channel. Normally, this mediator meets privately with each party, to exchange facts and to summarise their positions, and then proceeds with the presenting of offers and counter-offers to the parties until an agreement is reached. The reason for the success of mediation is that parties are more likely to disclose classified information to the mediator than to their counterpart, as negotiations should be kept confidential, and the final decision is still up to the parties.

In recent years, in an attempt to avoid costly and time-consuming courtroom litigation procedures, the construction industry has encouraged the use of international commercial practices for solving all sort of disputes (Murdoch and Hughes, 2008). Arbitration is a formal dispute resolution procedure which can have many variations, depending on the specific construction legal systems: adjudication boards, dispute review boards, mini-trials, private judging and so on. The fundamentals of arbitration practice are that, although no one is compelled to arbitrate unless agreed to do so, once it is accepted, the final decision is binding and can only be reviewed by a court of law if there is a procedural malfunction, fraud or conflict of interest can be proved. However, there are systems where the decision can be submitted to a court of appeal. The panel of arbitrators, normally three in number, one named by each party, and the other by consensus between the other two, should be impartial and should have standard qualifications and minimum training requirements (for instance, ten years of construction industry experience). During the process, arbitrators can call witnesses, experts and the parties themselves to testify, together with other formal courtroom procedures. Besides being less costly and time-consuming than a court of law, the advantages of arbitration are that arbitrators are recognised experts in the construction field, familiar with industry practices (unlike most judges), so they are concerned that both parties feel that they reach an equitable resolution of the dispute.

As referred to above, there are some alternative dispute resolution methods, such as the Dispute Adjudication Boards provided by the FIDIC General Conditions (FIDIC, 1999), or the Dispute Review Boards in some countries. These methods advocate a problem-solving approach instead of a more formal and legal approach, and can also include: conciliation, structured negotiation, non-binding expert appraisal, expert determination and mini-trial. (Cheung, 1999). The main difference from the arbitration process is that the boards are constituted and agreed for each construction contract, at the commencement date, and regardless of the occurrence of a claim event. As with arbitrators, the members of the boards should be construction industry experts recognised by both parties, acting impartially and objectively, whose main assignment is to get information about project progress and observe construction problems as they occur, being able to encourage parties to deal with them promptly and to realistically cooperate with each other.

If there is no agreement to use arbitration, either established in the contract or subsequently arranged between the parties, the way is to go to litigation, using the conventional legal system of the jurisdiction relevant to the specific construction contract. In this situation, formal and rigid procedures apply, and the court of law decides the merit of the case, in terms of legal evidence, sustained facts and damages evaluation (Moura and Teixeira, 2010). However, parties to the contract should use their utmost efforts to avoid claims escalation, by using claim avoidance practices.

14.7   Claim avoidance practices

In order to minimise the risk of costs growing due to the mismanagement of the claims process, construction industry practitioners developed a series of techniques and recommendations to improve claim reduction during the construction phase. These techniques are essentially focused on improving the quality of contract documents, as in most cases, construction claims result from problems originating in the design or in the pre-construction phase. Independent design quality review, mandatory insurance against errors and omissions or the demand for total quality management of architectural and engineering firms, are some of the recognised measures to ensure the quality of contract documents and avoid construction claims.

One common source of claims and disputes, representing a significant risk of cost and time overruns, is the different site conditions encountered by contractors. This situation results in most of the cases, in insufficient geotechnical subsurface studies or errors in data interpretation. For that reason, a new concept is introduced especially in underground construction projects or in projects where the most accurate definitions of subsurface conditions are necessary, that is geotechnical baseline reports (Rubin et al, 1999). Even though it is impossible or at least economically infeasible to determine what the exact subsurface conditions are, these reports aim to present the designer's description and interpretation of the site surveys and boring logs made on behalf of the project. Furthermore, the reports should anticipate subsurface behaviour in respect to the most likely construction method to be used by contractor, as well the necessary information about slope stability, dewatering methods or the strength of bedrock levels.

Another technique sometimes used to prevent and solve contract disputes is the escrow of bid documents (Rubin et al, 1999). This process consists of a controlled disclosure by the contractor of the documents used to support the bid, such as worksheets, cost estimation and vendor proposals, which can be put in escrow for future use, if and when a claim arises. These documents that belong to the contractor should be sealed and presented at the time of the bid, remaining in a safe or in the possession of a third party, as the information is confidential and considered a trade secret. If some controversy takes place, then the seal is broken with the agreement of both parts. In some cases the mere existence of the documents can avoid disputes.

Another type of measure that can be implemented at the design or pre-construction phase is the constructability review and value engineering incentive clauses. Constructability can be defined as the optimum use of construction knowledge and experience in planning, design, procurement and field operations, to achieve overall project objectives. Normally the constructability reviews are done in the final stage of design process, by the construction staff experienced in past projects and familiar with claims and disputes presented in those projects (Fisk and Reynolds, 2009).

Value engineering is the assurance that the built facility is adequate for its function at the lowest reasonable life-cycle cost, where the value index refers to the ratio of the worth of materials or methods required to provide the function, against their cost (see Chapter 12 for more information). This approach can be used either at the design phase, where the implementation can provide higher savings, or at the construction phase through an incentive clause in the contract, where the savings from value engineering studies are apportioned between client and contractor. Normally, this contract clause motivates the presentation of more cost-effective solutions, while still meeting the project's objectives, because it guarantees rewards for discovering value engineering enhancements.

The final measure that deserves to be mentioned is the partnering approach, which can be defined as the establishment of a working team among the parties, for mutually beneficial resolution of the ongoing difficulties and problems that typically arise on a construction project (see Chapter 3 for details). The objective of this technique is to set up a climate of cooperation, communication, fair play and mutual confidence between client, contractor, designer and all the other stakeholders, internal or external to the construction project, and this can start even before the bid auction and the tender stage. This process consists of voluntary workshops, seminars and meetings that help the parties to establish working relationships in a non-adversarial atmosphere, where arising problems can be discussed and resolved, avoiding future formal claims.

14.8   Management of the change process

Change is inevitable, especially in construction projects, owing to their uniqueness, unpredictability or limited resources spent at the design phase. Normally, every construction contract includes change clauses, authorising the client to order the contractor to do something different from what was expected, namely any additions, deletions or substitution of work, or other revision to the project goals and scope. Given that a change order can cause substantial adjustments to the contract duration and cost (Pinnel, 1998), the occurrence of changes in construction should be properly managed, in order to anticipate potential problems, thus reducing the disruptive effects of unpredicted change.

Problems caused by changes taking place in construction can be classified in several ways. According to the consequences or severity to the project, a change order can be classified as: a gradual change, if it happens during a prolonged period with low intensity; or a radical change. Changes can also be classified as: anticipated, if they are planned in advance and occur as planned; or as emergent if they arise spontaneously and could not have been anticipated. In respect to its necessity, construction changes can be classified as: required, if there is no option but to make the change; or as elective if it is not mandatory and one can choose whether or not to implement the change.

There are many different causes of change, and they can be classified into external or internal causes, occurring either in the design or the construction phases. Some examples of change causes are: design errors, design improvements, different site conditions, ambiguities in project goals or scope, errors or conflicts in specifications, inadequate materials, regulatory issues, new mandatory specifications, regulations and lawsuits. While some changes may bring benefits (for instance, when involving technological advances, environment issues or value engineering principles), most of them, if not properly managed, can result in severe cost and time overruns, including the possibility of demolition and rework.

For that reason some authors (Sun, 2005; Ibbs et al., 2001) have developed change management systems to be used by project teams, with the aim of responding quickly to the occurrence of a change, facilitating contingency plans for unanticipated changes, and avoiding costly uncontrolled effects. These processes are required to ensure that each individual and their organisations put themselves in the best position to cope with the changes that occur. The management of change does not start when the owner wants to issue a variation order or when a contractor receives a change order: it starts at the project's conception and continues until its completion.

Normally, these processes consist of four main steps: change identification, change evaluation, change approval and change implementation. The aim of the first step is to seek and identify areas where potential changes are likely to occur, in order to recognise them at an early stage and to prepare responses proactively, protecting the construction project from unexpected losses and costs overruns (Ibbs et al., 2001). Once changes are identified and recognised (either they are required or elective), the next step is change evaluation in order to determine whether the management team should accept and implement it (if elective) and to get the necessary funding for immediate approval, as any delay can increase costs. The approval stage is the following step, and it is required to obtain to project manager or client's authorisation to implement the change; sometimes it involves several iterations, in order to analyse other options or negotiate cost estimations. Finally, the implementation phase occurs when the modification is communicated to the field personnel whose work is affected by the change, and adjustments are made in the schedule of the construction works. The stages of the change management procedure, as proposed in literature, are presented in Figure 14.4.

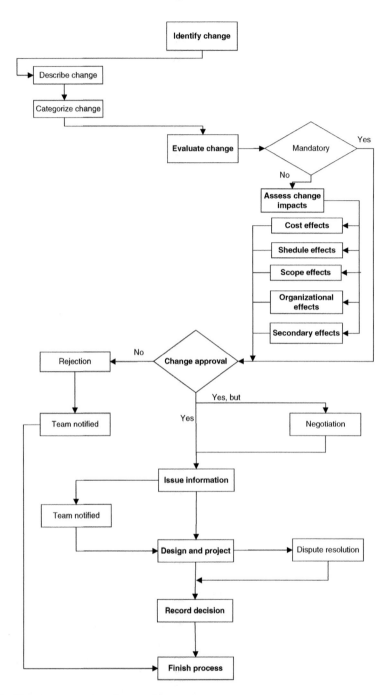

Figure 14.4   Change management procedure.

More disputes and claims arise from actual or perceived changes in a construction project than for any other reason, creating opposing positions: on the one side contractors that threaten work stoppages if they don't get paid for additional work resulting from changes, allegedly not considered in the original tender; on the other side are clients insisting that unknown conditions and extra features should be covered by the original bid price. So the implementation of change management in order to predict change whenever it is viable, and adequately reacting to change after its occurrence, will minimise the negative impact of inevitable changes in the normal progress of a construction project.

References

Adrian, J.A. (1993). *Construction Claims: A Quantitative Approach*. Stipes Pub. LLC, Champaign (Illinois).

Bramble, B.B. and Cippollini, M.D. (1995). *Resolution of Disputes to Avoid Construction Claims*. Transportation Research Board, Washington D.C.

Carnell, N. (2000). *Causation and Delay in Construction Disputes*. Blackwell, Oxford.

Cheung, S. (1999). Critical factors affecting the use of alternative dispute resolution processes in construction. *International Journal of Project Management*, 17(3), 189–194.

Davison, R. (2003). *Evaluating Contract Claims*. Blackwell, Oxford.

Diekmann., J. and Girard, J. (1995). Are contract disputes predictable? *Journal of Construction Engineering and Management*, 121(4), 355–363.

FIDIC (1999). *Conditions of Contract for Construction*. International Federation of Consulting Engineers, Geneva.

Fisk, E.R. and Reynolds, W.D. (2009). *Construction Project Administration* (9th Edition). Pearson, Upper Saddle River (New Jersey).

Horner, R.M.W. and Talhouni, B.T. (1996). *Effects of Accelerated Working, Delays and Disruption on Labour Productivity*. The Chartered Institute of Building, Ascot.

Ibbs, W., Wong, C. and Kwak, Y. (2001). Project change management system. *Journal of Management in Engineering*, 17(3), 159–164.

Kululanga, G., Kuotcha, W., McCaffer, R. and Edum-Fotwe, F. (2001). Construction contractors' claim process framework. *Journal of Construction Engineering and Management*, 127(4), 309–314.

Kumaraswamy, M.M. (1997). Conflicts, claims and disputes in construction. *Engineering Construction and Architectural Management*, 4(2), 95–111.

Kutner, S. (1989). Claims management. In: *Project Management: A Reference for Professionals* (Kimmons, R.L. and Loweree, J.H. eds.), Marcel Dekker Inc., New York.

Levin, P. (1998). *Construction Contract Claims, Changes & Dispute Resolution* (3rd Edition). ASCE Press, Reston (Virginia).

Moura, H. and Teixeira, J.C. (2007). Types of construction claims: a Portuguese survey. In: *Proceedings of XV Annual Conference of ARCOM* (Boyd, D. ed.), 129–136, Belfast.

Moura, H. and Teixeira, J.C. (2010). Managing stakeholders conflicts. In: *Construction Stakeholder Management* (Chinyio, E. and Olomolaiye, P., eds.), 286–316, Blackwell, Oxford.

Murdoch, J. and Hughes, W. (2008). *Construction Contracts: Law and Management* (4th Edition). Spon Press, London.

Pinnel, S. (1998). *How to Get Paid for Construction Claims: Preparation, Resolution Tools and Techniques*. McGraw-Hill, New York.

Powell-Smith, V., Stepheson, D. and Redmond, J. (1999). *Civil Engineering Claims* (3rd Edition). Blackwell, Oxford.

Rubin, R., Fairweather, V. and Guy, S. (1999). *Construction Claims: Prevention and Resolution* (3rd Edition). Wiley, New York.

Semple, C., Hartman, F. and Jergeas, G. (1994). Construction claims and disputes: causes and cost/time overruns. *Journal of Construction Engineering and Management*, 120(4), 785–795.

Sun, M. (2005) *Managing Changes in Construction Projects*. Engineering and Physical Science Research Council, Bristol.

Zaneldin, E., (2006). Construction claims in United Arab Emirates: types, causes and frequency. *International Journal of Project Management*, 24, 453–459.

Further reading

Hutchings, J. (1998). *Construction Claims Manual for Residential Contractors*. McGraw-Hill, New York.

Pickvance, K. (2010). *Delay and Disruption in Construction Contracts* (4th Edition). Sweet & Maxwell, London.

15 Project Closeout

15.1 Educational outcomes

The completion and closing process of the construction phase includes different activities: productive, regulatory, organisational, bureaucratic and preparatory, for the operation of the facility. The contractor and the owner must perform most of them jointly. By the end of this chapter, readers will be able to:

- analyse the completion and closing process of the construction phase of the facility life-cycle
- understand the regulatory tasks of testing and commissioning
- explain the bureaucratic tasks related to the handover, guarantee period and final payment
- describe the post-project review.

15.2 The closeout process

The completion of the construction project is the final activity developed by the construction company and the project manager (as representative of the owner), prior to the owner's acceptance and occupation of the facility. For most of the tasks at this stage, coordinated actions are required between the project manager and the contractor. The project management team is composed of technicians and advisors appointed by the owner, who are directly responsible for the inspection and verification of the correct execution of the works (CIOB, 2010). Within this group, the project manager (architect or civil engineer), the safety and health coordinator and the quantity surveyor are

Construction Management, First Edition. Eugenio Pellicer, Víctor Yepes, José C. Teixeira, Helder P. Moura and Joaquín Catalá.
© 2014 John Wiley & Sons, Ltd. Published 2014 by John Wiley & Sons, Ltd.

included; the latter is only needed in some countries such as the UK or Spain for building projects.

From the point of view of the contractor, the closing of the project is one of the most difficult activities of the construction phase, comprising a range of different tasks (Fisk and Reynolds, 2009). On the one hand, purely productive tasks must be completed before the deadline, in accordance with the technical specifications of the project. On the other hand, there are activities related to the internal organisation of the construction site team, compulsory tests and inspections, bureaucracy requested by the contract or the preparation of the facility's operation. The tasks related to the construction site team can be classified in three types: productive, dismantlement and management-based. Productive tasks are the usual ones in the execution of a construction project, with the added difficulty of requiring the completion of the project within the deadline. The tasks related to the dismantlement of the facilities on site and the management of the closeout process are analysed in the following section.

Tests and trials, as well as any inspection and verification activities carried out, are usually the responsibility of the contractor, following the project manager's guidelines. In this case, passing the tests and trials will entail the total or partial approval of the corresponding facility, section or element. If the contractor carries out tests and trials at his own expense and following internal guidelines, without the instructions of the project manager, he/she risks having to repeat them. In any case, these can be completed in compliance with the inspection activities that lie within the responsibility of the contractor (CIOB, 2010). Section 15.4 shows the tasks related to inspection and testing.

Bureaucratic tasks included in the contract can be summarised as follows: handover, guarantee period and final payment; all of them are described in section 15.5. The final processes that must be generated to finish the contract are explained in the following sections: as-built project, operation and maintenance manual and final report. Most of these tasks are carried out directly by the project manager (PM), with the collaboration of the contractor when needed. Figure 15.1 shows a draft of the process, for orientation purposes, since it can vary, in accordance with the country, type of works, owner and contract.

15.3   Completion and closing of the construction project

When the construction project enters the final phase, pending work can exceed the estimated deadline and budget. In addition, until the last moment, the construction site team can be too optimistic and underestimate the pending workload. Additional risks appear during this phase, such as the lack of final coordination, the non-compliance with the owner's requirements, potential omissions or errors, additional work that exceeds the contract's requirements, the tendency to perfect the works for an indefinite period of time, or the stress during the final weeks of the project, among others

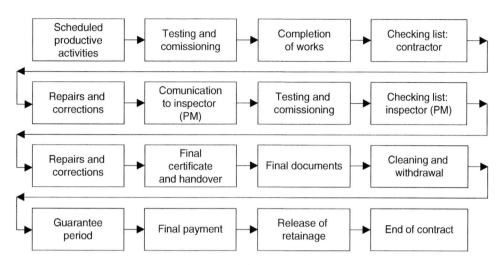

Figure 15.1 Layout of the closing process (developed from CIOB, 2010).

Code	Works unit	Company representative	Subcontractor	Dates			Observations
			(in case required)	Estimated start	Estimated end	Actual end	

Figure 15.2 Checklist.

(Nicholas and Steyn, 2008; Turner, 2008). The construction site manager must meet with the team to reduce the risks during this final stage.

The end of conventional productive tasks entails the completion of many different details. These can involve the execution of the appropriate tests, directly or in the collaboration with the project manager. Check lists (or 'punch' lists) can be helpful during the systematic revision of the construction project. This verification is recommended before reporting the completion of the works to the project manager (Mincks and Johnston, 2010). The results of this verification will include the list of defects and faults that must be solved as soon as possible, as well as the parties responsible for correcting them and the time available to carry them out (Figure 15.2). Therefore, the execution of final details, terminations and other activities in works units that have not been correctly executed or ignored in some cases, is recommended. The replacement of faulty materials or equipment might be necessary (Fisk and Reynolds, 2009). In this case, the supply of materials or equipment might present the risk of delaying the end of the construction project. In the case of the intervention of subcontractors in the repairs, special monitoring activities are required to avoid future claims or problems in meeting the deadline. In any case, the use of protocols and procedures for the completion of the construction project can optimise the results obtained.

The drafting and management of verification lists might entail an excessive waste of time for the technicians involved in the project completion and closing process. The best way in which this can be achieved is with continuous inspection throughout the construction process in order to minimise the extension of elements or parts of the works that must be repaired or completed later. This approach is valid for both the internal inspection of the construction company and that carried out by the project manager on behalf of the owner. General checklists can be used by the company for this type of works, based on previous experience. The computerisation of the process can be achieved with the use of a management system based on a database, accessible from the company's intranet, laptops and personal digital assistants (PDAs); this system can be very useful for recording anomalies and their subsequent tracking and analysis (Tserng et al., 2005).

The contractor has to complete and finish the project, ensuring that the optimum usage conditions are present; there must be no visible signs of the presence of the company or the execution of the works. Therefore, some tasks are fundamental for the dismantling of the company's jobsite facilities and, in addition, allowing the maintenance of the facility during the guarantee period, if stated as such in the contract. The activities that must be carried out during this phase by the construction team are listed in Box 15.1.

Box 15.1   Activities developed during the completion of the project

- meeting with the owner to confirm that the construction project was completed in accordance with the specifications, measuring the degree of satisfaction with the activities carried out
- meeting with the team to assess the development of the construction project
- adequate ordering of the documentation generated (photographs, videos, letters, reports, etc.) and selection of what must be stored from the handover on
- cleaning of the site and recovery of the works facilities, including:
  - reallocation of the specific productive equipment (machinery, vehicles, etc.)
  - dismantlement of the jobsite facilities and services (offices, communications, computers, furniture, vehicles, etc.)
  - removal or demolition of auxiliary facilities (parking spaces, supplies, etc.) and any visible element that indicates the existence of the construction works (fences, signs, etc.)
  - transfer of the construction site management team to other destinations, ensuring the transition is as smooth as possible

---

(Continued)

> o burial or concealing of elements that must be hidden (foundations), whenever the terrain enables such a task
> o the execution of the restoration plan (including landscaping if needed), in the case of a quarry, gravel site or landfills
> o removal of waste at the site (dangerous, inert and urban) and their adequate management, in accordance with the local legislation and contract regulations.
> - managing paperwork, licences and authorisations required (administrative, labour, industrial, tax, environmental, etc.)
> - notification to the project manager (as representative of the owner) that the construction project is finished, after checking the level of completion and development of the works, with the purpose of ensuring correct handover
> - reimbursement of warranties presented by the company at the signature of the contract
> - economic settlement of the contract.

In addition, the documentation accumulated by all of the personnel participating in the project must be ordered: reports, studies, meeting minutes, modification orders, offers and contracts, invoices, letters, construction charts, technical specifications, calculations, photographs, videos, etc. The administrative department of the company selects the useful information; later, it is stored, while the remaining documentation is destroyed. This cleaning and storage phase also includes digital archives, so that final information is stored on CD-ROM or DVD disks; the documents with original signatures must be stored on paper, for a specific period of time, in accordance with the legal consequences that could be derived from not filing that information (Mincks and Johnston, 2010). In any case, the selection process must be carried out after the as-built project, maintenance and operations manual and final report are drafted, so that these documents are also included in the previously filed information (CIOB, 2010).

15.4 Inspection and tests

The activities described in this section do not only take place during the final construction phase of the facility life-cycle, but rather from the start of the project. In any case, the perfect execution and operation of the facility can only be verified during the final stage, so that its importance grows throughout the execution of the project (Turner, 2008). The verification and inspection tasks, in each element and as a whole, under different situations (for example, weather), are very important. In general, facilities (both in building and civil engineering) require a greater intensity of inspection, especially in the case of mechanical and electrical equipment.

This section has taken a building project as reference for the text, following the model set forth by the Chartered Institute of Building (CIOB, 2010). In general, this model can be applied, in simplified form, to civil engineering works and to other countries. As stated above, the tests and trials can be carried out by the contractor, at his/her own expense or under the supervision of the project manager; the latter can also carry out these activities alone. Depending on the circumstances, these tests and trials can be classified in four categories:

- static
- dynamic
- required by external bodies
- operational.

Static tests are carried out throughout the execution of the works to verify the application of good construction practices and the quality of the works that must be examined. Dynamic tests are used to guarantee that the services/ units/works sections operate in accordance with the forecast design and technical specifications. In some cases, the inspection can lead to the need to obtain a certificate; examples include pressure and water-tightness tests in water or gas pipes. The inspections carried out by external organisations can generate partial building occupation certificates. It is also necessary to check the correct functioning of the facility as a whole. In the case of buildings, the pertinent environmental verifications are also required, which might involve the execution of different tests, according to the time of the year, even after the client has occupied the building. The main inspection tasks to be carried out by the project management team are described in Box 15.2 (CIOB, 2010).

Box 15.2   Inspection tasks

- programme all tests and trials required, in coordination with the contractor
- inspect the work and inform on the compliance with technical specifications, stating the corrective actions required
- measure progress using specific tests
- check that all compulsory documentation is available at any given time
- coordinate the necessary tests with external organisations and insurance companies
- record the tests carried out:
  - date and location
  - facility or element tested
  - regulations and techniques applied
  - inspector
  - result.
- obtain operating manuals with the corresponding documentation (in paper and digital files), including the computer programs required.

15.5   Handover

The completion and delivery of the facility are closely linked. It is the final stage of the activities carried out by the contractor and the project manager, prior to the acceptance of the facility by the owner. This is usually carried out in coordination between both parties. First the contractor must inform the project manager in written form of the date estimated for the completion of the construction project, in order to hand over the facility. This is one of the main tasks that must be carried out by the project manager, which will include the presence of the owner and the contractor. It creates the document called the final certificate or certificate of completion (Fisk and Reynolds, 2009).

If the construction site is in perfect condition, in accordance with the specifications, the project manager appointed by the owner authorises the correct reception of the project, issuing the corresponding certificate of completion. When the construction project is not in perfect condition, this is stated on the certificate of completion and the project manager states the defects detected, describing the precise instructions and fixing a period so that the contractor can solve those defects. If after that period the defects detected were not solved, an extension could be given to the contractor, which must be complied with, or the contract could be terminated. In any case, partial final certificates could be signed. The signature of the certificate of completion is a turning point in the process, since it transfers the responsibility of the facility from the contractor to the owner. The project manager must ensure that the owner has the adequate guarantees and safety measures necessary to take this step (CIOB, 2010).

On signing the certificate of completion, the guarantee period starts, with no sort of modification or interpretation valid after that signature. The guarantee period is established in the contract, in accordance with the nature and complexity of the project. Guarantees used from the start of the project can be maintained until the end of the guarantee period. These instruments are mainly used in case a subsequent claim is made, since they guarantee the owner a minimum economic compensation (Fisk and Reynolds, 2009). During the guarantee period, the contractor may be in charge of the maintenance, in accordance with the specifications and instructions provided by the project manager, if it is established this way in the contract. During the pre-established period, prior to the fulfilment of the guarantee period, the project manager drafts a report about the state of the facility. If it is favourable, the contractor is exempted from any responsibility, proceeding to the reimbursement or cancellation of the guarantee and corresponding settlement of pending obligations. When the report is not favourable and the defects observed are due to deficiencies in the execution of the works and not due to the normal use of the facility, during the guarantee period, the project manager establishes the appropriate instructions for the contractor to implement the necessary repairs. A period is extended, during which the contractor continues to be in charge of the maintenance, with no rights to receive additional money for the expansion in the guarantee period.

After the expiry of the guarantee period, if the report of the project manager on the state of the project is favourable, the project manager drafts a settlement proposal for the works executed, taking the economic conditions established in the contract as the base for the assessment (CIOB, 2010). The settlement process is notified to the contractor to receive its conformity. Within the period established for such purposes, the owner must provide the settlement and satisfy the resulting balance, whenever appropriate. Therefore, the settlement balance is the difference between the total cash amount and the amount valued at origin.

15.6 Occupation

The purpose of the occupation stage is to ensure that the facility is equipped and in operation, in accordance with the initial plans, in order to start with the facility's operation phase. The project manager (or the owner) must take into consideration the steps indicated in Box 15.3 (CIOB, 2010).

Box 15.3   Project manager's tasks during the occupation stage

- forecast occupation and operation objectives with enough time in advance
- install basic systems for the running of the facility: technology, information, storage, logistics, personnel training and so on
- obtain the permits and certificates required for the commissioning of the facility
- detail the spaces available and their occupation
- plan the migration (in the case of buildings):
  - define the building's occupation
  - establish the moving and material transfer deadlines
  - identify key activities and appoint supervisors
  - define the sequence of movements to minimise its influence on the business
  - identify potential risks that could affect the migration
  - inform the users
  - manage the movement and transfer of materials, including the corresponding contracts and tactical movements.
- manage the contracts with companies that carry out ordinary tasks in the operation phase: reception, safety, cleaning, maintenance, ICT, waste, gardening, courier services and so on.

The continuous maintenance of a facility requires a prior analysis to identify the works that must be carried out systematically. This analysis can lead to a project that defines the facility precisely as-built. A maintenance and operation manual can also be drafted with the list of preventive actions and procedures that must be taken into account during the operation of the facility. An example includes the systematic white line painting actions or the cleaning of drainage works on roads. When the facility has the mechanical or electrical equipment required (for example, a building, a dam or a wastewater treatment plant), these systematic activities are necessary (Mincks and Johnston, 2010).

The owner is not only responsible for the design and construction of the facility, but also for providing the best usage conditions possible to users after its handover. From the moment when it is planned, the owner must consider that the whole complex process carried out until the construction of the facility is required to satisfy the user at minimum cost. Therefore, the operation phase, from the owner's point of view, has two basic objectives: providing service to users, and preserving the durability and exploitation of the facility (Fisk and Reynolds, 2009).

The operation phase of the facility life-cycle, in the case of a building, has to deal with the management of the facility as a business, including maintenance activities. The management of the facility includes the economic, financing and administrative activities that involve the final users of the building. Maintenance looks to ensure the structural and functional performance of all elements of the facility in terms of stability, safety and liveability. Therefore, maintenance has to be attained at periodic intervals. Nonetheless, four types of occupation can occur during the operation phase:

- sale to user (home buyer): the title of property is passed on to the buyer, and the owner is relieved of its responsibilities (except for hidden defects)
- rent to user (home lender or tenant): the title of property is still kept by the owner. It cedes only the temporary use to the tenant, recovering the use once the contract is terminated
- operation by another company: the owner subcontracts the operation of the building (or part of it) to a specialised company; it can take the legal form of facility management or concession
- use by the owner: it may also include a transition time before a final decision for the definitive use of the facility has been set up.

15.7 Final documentation

The correct maintenance and operation of any facility requires an in-depth understanding of its performance. This involves establishing all of its characteristics and obtaining an inventory of them, as well as the agenda for the status of the degree of conservation and/or deterioration present. The documentation requested by the owner to implement the operation phase is as follows (based on CIOB, 2010):

- occupational health and safety records
- as-built project or, at least, the final plans and specifications
- the operation and maintenance manual
- guarantees of manufacturers and suppliers (certificates, quality seals, etc.)
- copies of minutes, certificates and permits
- documents related to tests and inspections
- final report.

Aspects relevant to the as-built project, operation and maintenance manual and final report are described next. In case of drafting an as-built project, the documentation structure will follow the guidelines set out in the original design project, with the following documents: summary report, calculations, drawings, technical conditions and final budget. The exact location and dimensions of services, facilities, elements and work units must be adequately stated. It is very important to explain the incidents occurred, changes implemented to the original design project and their justification (including the appropriate calculations). It is also convenient to include the documentation related to contractors and third parties (minutes, certificates, official correspondence, etc.). Finally, the budget must be detailed, comparing it with that initially estimated and including the variations in measurements and prices (Mincks and Johnston, 2010).

Many facilities make important investments in the production of their operation and maintenance manual. In many cases, manuals are not used and the effort made is wasted. The operation manual includes all information required to deliver the maintenance and operation characteristics of the facility to the service manager in a format that is easy to use. An effective manual requires the organisation of the manual, a content that is easy to read and understand, as well as graphs and charts that can help operators in their work. Therefore, the manual includes many layouts and photographs and the computer support required for their use (Mincks and Johnston, 2010).

There are many different options that can be followed for structuring the documentation, in accordance with the type of the facility, current legislation and criteria of the owner. The owner may require that the documentation be delivered not only on paper but also digitally. Generally, the files are saved using the portable document format (Adobe™ Acrobat™). Some owners may also require the 'source' files (such as *.dwg for drawings or *.doc or *.docx for documents); in the case of some contracts and local regulations, this fact may conflict with safeguarding the intellectual property of the author (as commented in Chapter 4 with the implementation of BIM). Nonetheless, Box 15.4 indicates a tentative table of contents for the final documentation.

> **Box 15.4   Table of contents for the final documentation**
>
> - summary report:
>   - project and background data
>   - companies and technicians participating in the execution of the works
>   - description of the project
>   - services affected
>   - acquisition of land
>   - description of materials, equipment and construction techniques used, including catalogues, certificates, seals and so on
>   - quality control executed during the construction phase
>   - quality control forecast during the operation phase
>   - main events detected during the construction phase, from the technical point of view.
> - drawings:
>   - general floor plan
>   - profiles and vertical drawings
>   - drainage
>   - accesses
>   - services
>   - installations.

It is also advisable to plan training sessions for the operation and maintenance of the facility, usually carried out by the supplier of equipment and aimed at the personnel who will carry out the operating and maintenance tasks (CIOB, 2010). These sessions must be duly documented too.

15.8   Post-project review

Companies in the construction industry produce and manage by projects (Gann and Salter, 2000). Projects are time-limited, whereas the team is specific for each construction project. Thus, at the end of the project team its members are dispersed, being assigned to other projects. In this environment, the transmission of knowledge is difficult and complex, and there is a serious risk that the knowledge acquired by individual members of the team may not revert to the company. The company must be able to capture the knowledge of its professionals so that it is useful for the entire organisation (Gann and Salter, 2000). Lessons learned from a finished project are usually too late to benefit this facility, but are potentially very helpful when applied to the design and construction of new facilities (Kartman, 1996).

The post-project (or postmortem) review is useful for both the owner and the contractor. Thus, this review can take two different approaches: 1. The owner's perceptions of performance by those parties employed by the owner in the project; and 2. The main contractor's internal evaluation of its performance (Griffith and Watson, 2004). The most common techniques used in order to capture the lessons learned from a project are (adapted from Griffith and Watson, 2004; Nicholas and Steyn, 2008): 1. Quantitative performance indicators, 2. Qualitative evaluation forms, 3. Meetings, 4. Workshops or conferences, and 5. Reports. A project audit can combine some of these techniques; for example, meetings with the different stakeholders and workshops can be a previous step to the final report, which can also include quantitative and qualitative indicators.

Whatever the technique used, it is of great importance to store the information correctly, using relational databases, and to manage this information through a database management system, linked to the internet or to the intranet of the organisation (Dikmen et al., 2008). It is important that the content be attractive to potential users. Furthermore, the more automated the system (including the capture of information) the more useful it will be. The process should also be standardised step by step, indicating the responsibilities of the various parties. As commented in Chapter 4, this system provides the organisation with centralised data control that allows sharing data and improvement of team and organisations interaction and integration.

Regarding quantitative and qualitative indicators, the first ones can be obtained through questionnaires that evaluate the project, using Likert scales for example, whereas the latter should be facilitated by the control system of the organisation. Post-project evaluation meetings look to improve processes and procedures, identifying aspects of good (and bad) practice and whatever other matters that may have been of concern during the project among the parties involved. Depending on the approach of the review, different stakeholders must be invited to the meeting; anyway, they should be the ones that work throughout the project or are affected by the final product (the facility), such as the end users. These meetings also allow the reinforcement of interpersonal and organisational relationships (Griffith and Watson, 2004).

Close-out (post-construction) workshops or conferences extend the assistance to a greater number of guests. The participants are usually key personnel in the organisation and relevant members of the project. Its goal is to recap lessons learned on the project in order to avoid mistakes and promote innovations. However, the workshops, as well as the meetings, are not useful if the knowledge acquired is only primarily transferred verbally among the attendees, so the minutes of the meeting or the summary of the workshop has to be made available to the whole organisation in a uniform and specific manner, and be easily retrievable for future use (Kartman, 1998). Sometimes, these lessons are not accepted by the rest of the organisation because of the 'not invented here' syndrome, in spite of being well documented and stored (Schindler and Eppler, 2003).

The final report records the project from different perspectives: contract documents and claims, management (scheduling, budgeting and organisation chart), progress payments, procedures, as-built design, changes (causes and consequences), history of the project, internal and external documentation (including storage and access) and any other interesting data for future use (quantitative or qualitative). Anyway, the purpose of the report is to state guidelines for the future, drawing together the positive and negative aspects (strengths and weaknesses) of the project as well as their causes (Dingsøyr, 2005). This way, the report identifies the methods and procedures that can be replicated or the ones that should be amended (O'Dell and Grayson, 2011). Likewise, the report can accumulate historical data that enable the subsequent analysis of similar projects. A report proposal based on the recommendations of the Chartered Institute of Building (CIOB, 2010) is detailed in Box 15.5 (CIOB, 2010).

Box 15.5   Main contents of the post-completion report

- project auditing:
    - brief description of objectives
    - summary of the amendments carried out in the original project
    - type of contract and contract conditions, stating their suitability
    - organisational structure, effectiveness and suitability of skills and experiences
    - scheduling of planned activities and milestones versus actual achievements
    - unforeseen issues and unusual events detected, as well as the solution adopted
    - brief summary of the strengths, weaknesses and lessons learned, with an overview of how the project was executed as regards the cost requirements, deadline, quality, technology, environment and health and safety
    - proposals for improvements in future projects.
- study of cost and time:
    - effectiveness of cost and budgetary controls
    - effectiveness of the claim procedures
    - planned versus actual cost, as well as original versus final budget
    - effectiveness of time control
    - planned versus actual schedule
    - impact of claims
    - identification of time extensions and additional costs due to amendments to original requirements and other reasons
    - analysis of initial and final scheduling, including the date established to complete the project and actual ending date, giving the reasons for variations.

(Continued)

- human resources:
  - communications and documentation channels (bottlenecks and their causes)
  - problems with subcontractors and suppliers
  - general assessment and comments on workforce health, morale and motivation.
- performance study:
  - planning and scheduling activities
  - validity of procedures used and controls established
  - resources to complete the project adequately in time and quality
  - identification of tasks carried out with or without success
  - assessment of the performance of subcontractors and suppliers.

Box 15.6   Barriers to knowledge transfer using post-project reviews

- team members' time constraints to document the lessons learned
- team members' denying bad performance (mistakes or low production) due to inadequate organisational culture
- deficient management support
- unfit project management, information and communications systems
- lack of incentives for the team members
- project-based nature that inhibits learning in spite of the fact that many processes are similar
- centralisation of knowledge capture within the organisation so diffusion is controlled.

The main barriers to knowledge transfer using post-project reviews are summarised in Box 15.6 (adapted from Dikmen et al., 2008 and Winch, 2010). The first barrier highlighted is generally concurrent with the others: team members do not have time to document the lessons learned. They are very busy trying to finish the current project; they can be mentally exhausted or even thinking about the next project. Furthermore, whatever work is not directly related to the project deliverables is not considered an added-value activity: it tends not to be done properly or even to be abandoned (Winch, 2010). Many times, the project manager is not aware of this issue or he/she does not consider it as his/her responsibility; thus, there is no proper planning of the closure of the project. However, this is a characteristic of the construction industry, where generally multiple projects are developed at the same time in each organisation. In the software sector, the post-project review is taken very seriously; Dingsøyr (2005) reveals that companies such Apple and Microsoft produce postmortem documents between 10 and 100 pages each, using specific teams for this purpose during three to six months.

References

CIOB (2010). *Code of Practice for Project Management for Construction and Development* (4th Edition). Wiley-Blackwell, Oxford.

Dikmen, I., Birgonula, M.T., Anac, C., Tah, J.H.M. and Aouad, G. (2008). Learning from risks: A tool for post-project risk assessment. *Automation in Construction*, 18(1), 42–50.

Dingsøyr, T. (2005). Postmortem reviews: purpose and approaches in software engineering. *Information and Software Technology*, 47(5), 293–303.

Fisk, E.R. and Reynolds, W.D. (2009). *Construction Project Administration* (9th Edition). Pearson, Upper Saddle River (New Jersey).

Gann, D.M. and Salter, A.J. (2000). Innovation in project-based, service-enhanced firms: the construction of complex products and systems. *Research Policy*, 29, 955–972.

Griffith, A. and Watson, P. (2004). *Construction Management. Principles and Practices.* Palgrave MacMillan, London.

Mincks, W.R. and Johnston, H. (2010). *Construction Jobsite Management* (3rd Edition). Thomson, New York.

Kartam, N. (1996). Making effective use of construction lessons learned in project life cycle. *Journal of Construction Engineering and Management*, 122(1), 14–21.

Nicholas, J.M. and Steyn, H. (2008). *Project Management for Business, Engineering, and Technology* (3rd Edition). Butterworth-Heinemann, Burlington (Massachusetts).

O'Dell, C. and Grayson, C.J. (2011). *If only We Knew What We Know.* The Free Press, New York.

Schindler, M. and Eppler, M.J, (2003). Harvesting project knowledge: a review of project learning methods and success factors. *International Journal of Project Management*, 21(3), 219–228.

Tserng, H.P., Dzeng, R.J., Lin, Y.C. and Lin, S.T. (2005). Mobile construction supply chain management using PDA and bar codes. *Computer-Aided Civil and Infrastructure Engineering*, 20, 242–264.

Turner, J.R. (2008). *The Handbook of Project-based Management: Leading Strategic Change in Organizations.* McGraw Hill, New York.

Winch, G.M. (2010). *Managing Construction Projects* (2nd Edition). Wiley, Oxford.

Further reading

CIOB (2010). *Code of Practice for Project Management for Construction and Development* (4th Edition). Wiley-Blackwell, Oxford.

Cotts, D.G., Roper, K.O. and Payant, R.P. (2009). *The Facility Management Handbook* (3rd Edition). AMACOM, New York.

Index

Construction Management, First Edition. Eugenio Pellicer, Víctor Yepes, José C. Teixeira,
Helder P. Moura and Joaquín Catalá.
© 2014 John Wiley & Sons, Ltd. Published 2014 by John Wiley & Sons, Ltd.

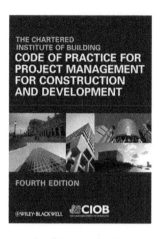
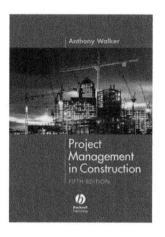
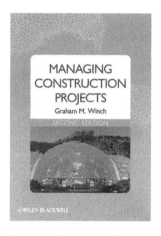

Printed and bound by CPI Group (UK) Ltd, Croydon, CR0 4YY

27/10/2024

14580305-0001